SpringerBriefs in History of Science and Technology

More information about this series at https://link.springer.com/bookseries/10085

Leo Corry

Distributivity-like Results in the Medieval Traditions of Euclid's Elements

Between Geometry and Arithmetic

 Springer

Leo Corry (ID)
History and Philosophy of Science
Tel-Aviv University
Ramat Aviv, Israel

ISSN 2211-4564   ISSN 2211-4572 (electronic)
SpringerBriefs in History of Science and Technology
ISBN 978-3-030-79678-5   ISBN 978-3-030-79679-2 (eBook)
https://doi.org/10.1007/978-3-030-79679-2

This Springer imprint is published by the registered company Springer Nature Switzerland AG
The registered company address is: Gewerbestrasse 11, 6330 Cham, Switzerland

Acknowledgements

As becomes evident from the secondary literature cited above, the background to my discussion on certain works of Islamicate mathematics relies strongly on the recent work of Jeffrey Oaks. My sincere thanks go to him for carefully reading parts of my text and for profusely commenting on it. He not only spared me some embarrassing mistakes but also helped me better understand some very important points about the views of the mathematicians whose works I discuss here, as well as of some of others whose work I do not address this time.

Concerning the Hebrew mathematical texts discussed above, I have benefitted from interchanges with Naomi Aradi, Stela Segev, Avinoam Baraness, and Ruth Glasner. Friendly thanks go to them for their patience and interest, and for insightful comments. I thank particularly Avinoam Baraness for allowing me to use the well-thought and well-designed diagrams that he prepared for the edition of Alfonso de Valladolid's text in his joint book with Ruth Glasner. The diagrams are published here by permission of Springer Nature. Special thanks go to Fabio Acerbi for sharing with me his thoughts and some important material on Byzantine mathematics.

Needless to say, all mistakes or inaccuracies remaining in the text are of my sole responsibility.

Contents

Chapter 1
Introduction

Abstract The present work explores the role and presence of distributivity-like properties in some texts of the medieval Euclidean traditions. It starts with an overview of propositions in Euclid's *Elements* which, retrospectively seen, embody treatments of distributivity-related properties of multiplication over addition or subtraction. It examines the way in which, taken together, these propositions and their interrelations are related to the all-important separation in Euclid's treatise between its arithmetic and geometric parts, and, consequently, between continuous and discrete magnitudes. The following main three sections discuss the treatment of distributivity-related situations in various medieval mathematical treatises, against the background of the significant changes that affected the interrelation between the two domains, geometry and arithmetic. The texts discussed belong to three main medieval mathematical traditions: Islamicate, Latin and Hebrew. They display a wide array of attitudes towards distributivity-like situations: some either explicitly or tacitly rely on the relevant Euclidean propositions, some formulate distributive like rules in order to attribute them a foundational role or to use them as ad-hoc resources in proofs, and some others simply bypass them even where their use would seem to be helpful. The perspective afforded by examining these distributivity-like situations, in various manifestations and in varying contexts, gives rise to fresh insights concerning medieval attitudes towards the questions of what are numbers and magnitudes, how they are used, what are their basic defining properties, and what is the right way to provide clear foundations for arithmetic as an autonomous field of mathematical knowledge.

Keywords Distributive-like rules · Euclidean tradition · Islamicate mathematics · Latin medieval mathematics · Hebrew mathematics

From the perspective of modern algebraic theories, in any system comprising two operations, the validity of a well-defined distributivity law typically embodies a kind of minimal requirement for the system to be mathematically interesting. Looking at premodern periods, however, it is rather curious to notice the shifting manners (and sometimes little attention or even total disregard), with which—in changing

L. Corry, *Distributivity-like Results in the Medieval Traditions of Euclid's Elements*,
SpringerBriefs in History of Science and Technology,
https://doi.org/10.1007/978-3-030-79679-2_1

historical contexts—focused interest was devoted to spell out the use of and the role played by distributivity-like properties in either geometry or arithmetic.

Consider for example the case, in the early modern period, of François Viète's *In artem analyticem isagoge* (1591). In defining the "Fundamental Rules of Equations and Proportions", on which "the science of correct discovery in mathematics" is to be based, Viète formulated sixteenth such rules. He stated that these "well-known fundamental rules … are given in the *Elements*", and that "analysis accepts them as proven". Among them we find the six "common notions", which indeed were explicitly formulated by Euclid in the opening section of Book I of the *Elements* (e.g., "(3) if equals are added to equals, the sums are equals"), but we also find additional statements that do not appear there. Some of these, or similar to them, appear scattered and implicit in later versions of the *Elements* or in later treatises in the Euclidean tradition. Thus for example: "(7) if equals are divided by equals. The quotients are equal", or "(10) if proportional are multiplied proportionally, the products are proportional".[1] But Viète's fundamental rules also comprise a distributivity rule that is not even hinted at in the *Elements* and, indeed, it is hard to determine whether it was formulated in this way, and attributed a foundational role, if at all, in any previous treatise of this tradition. Viète's rule (13) reads as follows (Viète 1983, 14):

> The [sum of the] products of several parts [of a whole] is equal to the product of the whole.

By looking at the way in which Viète, or any of his contemporaries, explicitly formulated this rule as a fundamental truth of mathematics, and then at the shifting ways in which similar ideas were previously handled in different mathematical cultures where the Euclidean text was studied, reedited, or relied upon, several interesting historical questions arise. I want to explore some of them in the present work.

The aim of this survey is to discuss ways in which "distributivity-like" situations arose and were addressed in various texts belonging to the Euclidean medieval traditions—Islamicate, Latin and Hebrew. Each of the examples examined below offers different perspectives on the issue. In a few of the texts considered we find something that comes close to a general, clearly formulated idea of "distributivity" and that is conceived as a fundamental, widely acknowledged and clearly defined, kind of property underlying the relationship between two basic and also abstractly defined operations, "product" and "addition" (like in the example of Viète above). In some places we find a direct and explicit reliance on one of the relevant Euclidean propositions while handling "distributive-like" situations, whereas in some others we find a tacit reliance on them. In some places they appear in a clearly geometric context, whereas in others we find explicit attempts to formulate a rule that is taken to be foundational for arithmetic. Some examples involve a clear separation between the handling of discrete and continuous magnitudes, whereas other examples mix both of them either intendedly or implicitly. Some other cases involve situations where

---

[1] For an exhaustive study of the development of the Euclidean axiomatics, see (Risi 2016). Viète's rules are not included in Risi's account.

a direct reliance on distributivity-like properties would seem to simplify the proofs and yet are not used.

The ideas discussed here in relation with "distributivity-like" properties developed and consolidated as part of a long process of interaction involving various kinds of ideas: product, area formation, addition, figure concatenation, numbers of various kinds, discrete and continuous magnitudes. I think that it is historically rewarding to look at these developments from a common perspective that involves a broad idea of "distributive-like" properties. Accordingly, then, the term is used here as a general, non-essentialist label that allows a common reference to various kinds of results that bear important similarities, rather than as an assertion that this was a clearly conceived, general idea specifically applied in particular cases.

I shall be referring here to some medieval texts already discussed in (Corry 2013). I will focus mainly on the mathematical issues directly connected with distributivity-like properties as such, and I shall not repeat here the historical descriptions and broader contextual considerations already discussed there. Those are highly important issues and they are also relevant to some of the questions discussed here. The interested reader is referred to (Corry 2013) for further details. For some of the treatises not discussed in my previous article I have added here some contextual information that is relevant to the main issues considered.

One important sense in which my analysis here is complementary to that of (Corry 2013) concerns the ways in which results from Book II of Euclid's *Elements* were treated by medieval authors, commentators and translators. I started by analyzing Euclid's own version of Book II, while fully endorsing the criticism put forward by Sabetai Unguru (Unguru 1975) vis-à-vis the so-called "geometric algebra" interpretation of Greek mathematics. At the same time, however, I stressed the inadequacy of an unqualified use of those terms (geometry, arithmetic, algebra) as if they referred to bodies of knowledge that are easily recognizable in ahistorical terms, and traces of which we may either clearly find or fail to find in any given ancient text. To the contrary, I discussed the ways in which new kinds of ideas gradually appeared in the medieval versions of Book II and transformed not only the formulations and the proofs of the propositions, but also the very borderlines and interactions among mathematical disciplinary matrices. It turns out that an analysis of the way in which distributivity-like properties appear in the medieval traditions sheds additional light on those same issues and provides original insights about medieval attitudes towards the questions of what are numbers and magnitudes, how they are used, what are their basic defining properties, and what is the right way to provide clear foundations for arithmetic as an autonomous field of mathematical knowledge.

I open the present analysis with a preliminary overview of some propositions of Euclid's *Elements* that, retrospectively seen, embody results that we can associate nowadays with a more general idea of distributivity. My intention is just to map out and to cursorily discuss a specific set of Euclidean propositions from books II, V, and VII, that, later on, *medieval authors* referred to or relied upon in their own works, when discussing the role of distributivity-like rules or applying them in specific situations. A main focus of interest will concern the way in which results and arguments that in Euclid's *Elements* appear in the clearly separate contexts of numbers, magnitudes and proportions, started to appear in medieval texts intermingled with one another. In the three subsequent sections I analyze, respectively, texts belonging to the three main medieval mathematical traditions: Islamicate, Latin and Hebrew. In another previous article (Corry 2016), I already discussed the cases of Jordanus and Campanus in relation with distributivity properties. Here I will only summarize the main points discussed there in greater detail, while connecting them to the other works that I analyze.

Throughout the text, I render in modern symbolic terms some of the results discussed. Unguru's (1975) article targeted such renderings as part of his criticism of the "geometric algebra" interpretation. In rendering parts of my presentation symbolically I have taken care to avoid any of the possible anachronistic interpretations that may misleadingly follow from doing so. No claim is made here that interpreting those results in modern algebraic terms is the right way to understand their content. In fact, besides this general disclaimer, which should apply to the entire text, I have added in some places indications about specific shortcomings of interpreting a given result symbolically. Still, in spite of these shortcomings the symbolic rendering sometimes serves as convenient shorthand to nail down a specific part in an argument or to cross-refer among results appearing the various texts. This is the reason to use them here.

References

Corry, Leo. 2013. Geometry and arithmetic in the medieval traditions of Euclid's *Elements*: A view from Book II. *Archive for History of Exact Sciences* 67 (6): 637–705. https://doi.org/10.1007/s00407-013-0121-5.

Corry, Leo. 2016. Some distributivity-like results in the medieval arithmetic of Jordanus Nemorarius and Campanus de Novara. *Historia Mathematica* 43 (3): 310–331. https://doi.org/10.1016/j.hm.2016.06.001.

De Risi, Vincenzo. 2016. The development of Euclidean axiomatics. *Archive for History of Exact Sciences* 70 (6): 591–676. https://doi.org/10.1007/s00407-015-0173-9.

Unguru, Sabetai. 1975. On the need to rewrite the history of Greek mathematics. *Archive for History of Exact Sciences* 15 (1): 67–114. https://doi.org/10.1007/BF00327233.

Viète, François. 1983. *The Analytic Art: nine studies in algebra, geometry, and trigonometry from the Opus Restitutae Mathematicae Analyseos, Seu, Algebrâ Novâ by François Viète*. Translated by T. Richard Witmer. Kent, Ohio: Kent State University.

Chapter 2
Distributivity-like Results in Euclid's *Elements*

Abstract The present chapter presents a preliminary overview of some propositions of Books II, V, and VII of Euclid's *Elements* that, retrospectively seen, embody results that we can associate nowadays with a more general idea of distributivity. It focuses on several results and arguments that are handled differently by Euclid in three clearly separate contexts: geometric magnitudes, arithmetic and numbers, and proportions. It also discusses Euclidean propositions where "visible" constructions implicitly use distributive-like properties.

Keywords Distributive-like rules · Euclid's *Elements* · Book II · Book V · Book VII · Eudoxian theory of proportions

In this chapter I present a preliminary overview of some propositions of Euclid's *Elements* that, retrospectively seen, embody results that we can associate nowadays with a more general idea of distributivity. As already stated, this overview is not intended as a detailed historical analysis of a general concept of distributivity that we can putatively attribute to Euclid. Nor do I wish to analyze here the actual, underlying conceptions behind Euclid's uses of additions and products of numbers or of magnitudes, and certainly not to elucidate questions related to their historical roots. Rather, my intention is just to map out and to cursorily discuss a specific set of Euclidean propositions from Books II, V, and VII. These three books of the *Elements* deal, respectively, with geometric magnitudes (lines and areas), proportions and numbers. A main focus of interest in the various chapters that follow concerns the way in which several results and arguments that I discus now—and that with Euclid were differently handled in these three clearly separate contexts—started to appear in medieval texts intermingled with one other, and either explicitly referred to or tacitly relied upon.

Fig. 2.1 Euclid's
Elements—II.1

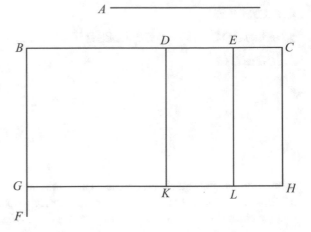

2.1   Book II

To the extent that distributivity-like properties are discussed as an issue of inherent, focused interest in Euclid's *Elements* that happens in Book II. Inasmuch as we can consider rectangle formation as a kind of multiplication, and the concatenation of rectangles having one side of equal length as addition, the first three propositions of Book II can be said to fundamentally embody distributivity-related ideas of the former over the latter. In a somewhat looser sense, but also very clearly, this statement applies also to propositions II.4,II.7. Proposition II.1, as is well known, formulates the following general property:[1]

> **II.1**: If there are two straight lines, and one of them is cut into any number of segments whatever, then the rectangle contained by the two straight lines equals the sum of the rectangles contained by the uncut straight line and each of the segments.

The main step in the argument of the proof relies on a proposition from Book I, I.34. In Fig. 2.1, side *BG* of the rectangle *BGHC* equals the given line *A*, and line *BC* is divided into sub-segments.

The said proposition I.34 is used to assert that, since *BK* is by construction a rectangle, then *DK* equals *BG* and hence equals *A*. A repeated application of this argument allows concatenating the three resulting rectangles into a single, larger one, and thus to complete the proof.

Propositions II.2–II.3 may be seen as particular cases of II.1:

> **II.2**: If a straight line is cut at random, then the sum of the rectangles contained by the whole and each of the segments equals the square on the whole (Fig. 2.2).

---

[1] All the quotations from the *Elements* from this section, as well as the accompanying diagrams, are taken from (Euclid 1956). For an enlightening discussion of the issues involved in the use of diagrams related to ancient Greek sources, see (Saito and Sidoli 2012).

Fig. 2.2 Euclid's
Elements—II.2

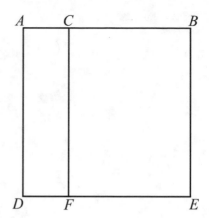

Fig. 2.3 Euclid's
Elements—II.3

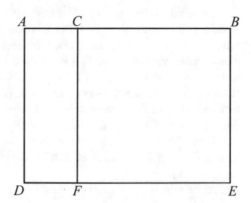

II.3: If a straight line is cut at random, then the rectangle contained by the whole and one of the segments equals the sum of the rectangle contained by the segments and the square on the aforesaid segment (Fig. 2.3).

In spite of their appearance as particular cases of II.1, Euclid did not prove them by straightforward application of the latter. Rather, the two proofs he offered essentially recapitulate the argument of II.1, but now applied to the particular cases in point. This reflects a more general feature typical of the proofs of the first ten propositions of Book II, namely that none of them relies on a previous one of the same book. Rather they are all proved by directly relying on propositions of Book I alone. In particular this is also the case of proposition II.4, which can as well be seen as a particular application of II.1. It embodies a distributivity-like property of square-formation over division of a line into two parts, which—like the previous two propositions—is not proved in the *Elements* based on II.1, but rather by direct application of I.34 and I.43. Its diagram is as in Fig. 2.4 and it is formulated as follows:

II.4: If a line is cut at random, then the square on the whole is equal to the squares on the segments and twice the rectangle contained by the segments.

Fig. 2.4 Euclid's
Elements—II.4

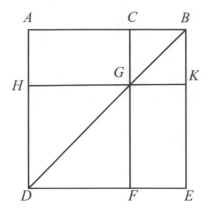

Proposition II.7 is quite similar in spirit to II.4, and the accompanying diagram is essentially identical. In fact, Book II comprises several propositions that can be arranged in pairs, which handle in a complementary way similar ideas, and in particular distributive-like ideas. That is the case with the pairs II.2–II.3 and II.5–II.6, as well as is with the pair II.4–II.7. In the second part of Book II, we have also the pair II.12–II.13, about which I say more below, but at this point it is interesting to stress that while II.4 provides the crucial, distributive-like argument for proving II.12, II.7 does the same for the proof of II.13. (Also, and although we will not delve into that here, II.5–II.6 provide in their own way the crucial arguments—that can also be described in this context as distributive-like—for the proofs of II.11 and II.14, respectively).

Proposition II.7 is stated as follows:

> **II.7**: If a line is cut at random, then the sum of the square on the whole and that on one of the segments equals twice the rectangle contained by the whole and the said segment plus the square on the remaining segment.

Thus, II.7 is quite similar in spirit to II.4, and its proof is likewise based on applying I.43 in a straightforward manner. In addition, the geometric situation embodied in their diagrams is essentially the same (Fig. 2.5).

2.2   Book V

The first six propositions of Book V can be seen as a self-contained, comprehensive discussion of results concerning equimultiplicity of continuous magnitudes (Acerbi 2003). As we will see below, some medieval treatises where distributivity-like properties are discussed do mix arguments taken from Book II with those taken from these specific propositions of Book V. For this reason I have chosen to include them here in this preliminary overview. Within the *Elements*, these propositions provide basic results that are used to develop in full the Eudoxian theory of ratios and proportions

Fig. 2.5 Euclid's
Elements—II.7

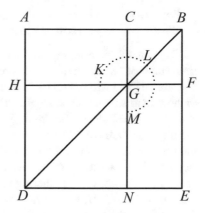

later in the book. Ratios and proportions as such, however, are not yet mentioned in these six propositions. They deal only with magnitudes and their multiplicities.

These are the enunciations of the six propositions:

V.1: If there be any number of magnitudes whatever which are, respectively, equimultiples of any magnitude equal in multitude, then, whatever multiple one of the magnitude is of one, that multiple also will all be of all.

V.2: If a first magnitude be the same multiple of a second that a third is of a fourth, and a fifth also be the same multiple of the second that a sixth is of the fourth, the sum of the first and fifth will also be the same multiple of the second that the sum of the third and sixth is of the fourth.

V.3: If a first magnitude be the same multiple of a second that a third is of a fourth, and if equimultiples be taken of the first and third, then also *ex aequali* the magnitudes taken will be equimultiples respectively, the one of the second and the other of the fourth.

V.4: If a first magnitude have to a second the same ratio as a third to a fourth, any equimultiples whatever of the first and third will also have the same ratio to any equimultiples whatever of the second and fourth respectively, taken in corresponding order.

V.5: If a magnitude be the same multiple of a magnitude that a part subtracted is of a part subtracted, the remainder will also be the same multiple of the remainder that the whole is of the whole.

V.6: If two magnitudes be equimultiples of two magnitudes, and any magnitudes subtracted from them are equimultiples of the same, the remainders also are either equal to the same or equimultiples of them.

Heath renders these propositions in modern algebraic symbolism, and his formulation sets the stage for discussing the possibility of interpreting them as general statements of distributivity laws. He renders them as follows:

V.1: $m \cdot a + m \cdot b + m \cdot c + \ldots = m \cdot (a + b + c + \ldots)$.

V.2: $m \cdot a + n \cdot a$ is the same multiple of a as $m \cdot b + n \cdot b$ is of b.

Further, says Heath, from the proof we learn that $m \cdot a + n \cdot a + p \cdot a + \ldots = (m + n + p + \ldots) \cdot a$.

V.3: $(m \cdot n) \cdot a = m \cdot (n \cdot a)$.

V.4: If $a{:}b :: c{:}d$, then $m \cdot a{:}n \cdot b :: m \cdot c{:}n \cdot d$.

V.5: $m \cdot a - m \cdot b = m \cdot (a - b)$.

V.6: If $n < m$, $m \cdot a - n \cdot a$ is the same multiple of a as $m \cdot b - n \cdot b$ is of b.

The general historiographical issues raised by the use of this kind of symbolic translation of Greek mathematics are well-known (Høyrup 2017; Schneider 2016; Unguru 1975). I will not rehearse that debate here. It is important, however, to focus on how, in this particular case, Heath's rendering is potentially misleading in a more specific sense. Indeed, the "multiple" of a given magnitude M is not, for Euclid, the outcome, $n \cdot M$, of a binary operation, namely, multiplying a given number n by the magnitude M. Nor is it the result, $M + M + M + M + \ldots + M$, of successive steps of binary additions of the magnitude to itself n times. Rather, it is more of an "accumulation" or of a "gathering together" of a *multitude* of instances of the said magnitude: $M, M, M, M, \ldots M$. Also, *V.2-V.3* above involve the particular difficulty that nothing like the coefficient multipliers m,n do appear in Euclid's text (Taisbak 1971, 65).

Heath's rendering also does not reflect (except in the case of V.4) the "if … then" style of formulating the propositions. Thus, a symbolic rendering that would remain closer to Euclid in at least this formal and important respect could be the following:

V.1′: If $a' = m \cdot a$, $b' = m \cdot b$, $c' = m \cdot c$,... then $a' + b' + c' \ldots = m \cdot (a + b + c \ldots)$,

V.2′: If $a = m \cdot b$, $c = m \cdot d$ and $e = n \cdot b$, $f = n \cdot d$, then, while $a + e = p \cdot b$, we also have $c + f = p \cdot d$. In this symbolic rendering one sees that $p = m + n$.

V.3′: If $a = m \cdot b$, $c = m \cdot d$ and $e = n \cdot a$, $f = n \cdot b$, then, $e = p \cdot a$, $f = p \cdot d$. In this symbolic rendering one sees that $p = m \cdot n$.

V.4′: If $a{:}b :: c{:}d$, then $m \cdot a{:}n \cdot b :: m \cdot c{:}n \cdot d$.

V.5′: If $a = m \cdot b$, $c = m \cdot d$, then $a - c = m \cdot (b - d)$.

V.6′: If $a = m \cdot b$, $c = m \cdot d$ and $e = n \cdot b$, $f = n \cdot d$, then while $a - e = p \cdot b$, we also have $c - f = p \cdot d$. In this symbolic rendering one sees that $p = m - n$.

Notice a point of particular interest in this regard concerning V.6. Euclid mentions separately the possibility that the remainders be "equal to the same". Symbolically this corresponds to the case $m = n + 1$ and there is no need to speak about it separately. But for Euclid, this is indeed a separate case, namely one in which we do not have a *multitude* of instances of the magnitude but just one instance. Hence the need to mention it separately.

The proofs of the first six propositions in Book V are based on various ways of counting the multitudes of the magnitudes involved in each case. Of special interest for us are the proofs of V.1 and V.2, because they will resurface in modified ways in the medieval texts discussed below. As for V.1, the proof is accompanied by a diagram as in Fig. 2.6.

Here AB and CD represent equimultiples of E, F respectively. The proposition states that AB, CD taken together is that same equimultiple of E, F taken together. The proof follows closely the diagram and comprises the following steps: AB is divided into (two) lesser magnitudes AG, GB (each equal to E) while CD is divided into the same number (two) of lesser magnitudes CH, DH (each equal to F). Now, the lesser magnitudes of the two kinds are joined into pairs (each equal to E, F) that

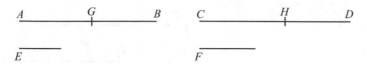

Fig. 2.6 Euclid's *Elements*—V.1

can be used to measure the sum of *AB, CD*. And the amount of such pairs turns out to be equal to the times (two) that each of the two lesser magnitudes measured both *AB* and *CD* at the beginning of the proof. In Euclid's words: "since *AG* is equal to *E* and *CH* to *F*, therefore *AG* is equal to *E*, and *AG, CH* to *E, F*. For the same reason *GB* is equal to *E*, and *GB, HD* to *E, F*. Therefore as many magnitudes as there are in *AB* equal to *E*, so many also are there in *AB, CD*, equal to *E, F*".

Once again for ease of reference below, I render the argument of the proof in modern symbols, while at the same time generalizing for more than two summands for the first and the second magnitudes. The steps of the argument are:

(a.1) $AB = E + E + E + \ldots + E$ (*n* times),
(a.2) $CD = F + F + F + \ldots + F$ (*n* times),
(a.3) $AB + CD = (E + E + E + \ldots + E) + (F + F + F + \ldots + F)$,
(a.4) $AB + CD = (E + F) + (E + F) + \ldots + (E + F)$ (*n* times).

Euclid lacked a flexible symbolic language that would allow him to formulate the proof in all of its generality. Still, the generality of the argument seems to be compromised neither in terms of the multiplicity involved ("two" in the proof) nor of the number of magnitudes that are added (here: *E* and *F*). In this particular sense, the generality implied in the symbolic rendering *V.1* does not seem to go beyond what Euclid stated in the enunciation and then proved. His main trick lies in pairing the measuring magnitudes and then counting the ensuing pairs.

Let us consider now the proof of V.2, whose diagram is reproduced in Fig. 2.7.

Here *AB* is the same multiple of *C* as *DE* is of *F*, and likewise, *BG* is the same multiple of *C* as *EH* is of *F*. The proposition states that *AG* is the same multiple of *C* as *DH* is of F. Notice that unlike in V.1, *C* and *F* need not be magnitudes of the same kind, and the proposition does not involve in any way gathering together the first or second magnitudes with the third or fourth, respectively. Rather, the proposition only compares two separate relationships between pairs, each pair comprising magnitudes of the same kind.

Fig. 2.7 Euclid's
Elements—V.2

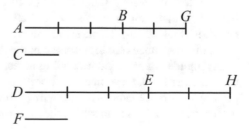

Fig. 2.8 Euclid's
Elements—V.5

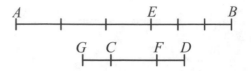

The starting point of the proof is that, as many magnitudes are in *AB* that are equal to *C*, so many are also in *DE* that are equal to *F* (and the same goes for *BG* and *EH*). It is a nice feat, I think, that Euclid attempted to infuse some kind of generality to the argument by taking three and two as the multiplicities (rather than, for example, two and two). Now the two multiplicities are gathered together with each other separately and as a result "as many as there are in the whole *AG* equal to *C*, so many are there in the whole *DH* equal to *F*."

The cases involving subtraction are considered in V.5 and V.6, which are parallel to V.1 and V.2. The proofs of the second pair are somewhat different from those of their counterparts in the first one. One point of interest concerning V.5 arises in reference to the accompanying diagram, which appears in Fig. 2.8.

Here we have,

(b.1)   *AE* is subtracted from *AB* to obtain *EB*, while *CF* is subtracted from *CD* to obtain *CF*;

(b.2)   the proposition then makes a statement about the remainder *EB*, namely that if $AB = m{\cdot}CD$, then $EB = m{\cdot}FD$.

(b.3)   Segment *CG*, not mentioned in the enunciation, is constructed as part of the proof, with the property that if $AE = n{\cdot}CF$, then $EB = n{\cdot}CG$.

The interesting point in the proof is Euclid's assumption of the existence of *CG*. Indeed this is tantamount to assuming that given any magnitude, here *EB*, we can divide it into as many equal parts as we wish. In Euclid's own diagram, for instance, *FD* is a third of *AE*, so that *CG* needs to be made a third of *EB*. This would seem to present no problem if we were dealing with segments, as in the diagram, but in fact it is only in VI.9, that Euclid proves that such a fourth line always exists.

In the case of areas or volumes, it would seem more difficult to figure out how to find out a fourth magnitude, parallel to *CG* in the above case. And, much worse, in the case of angles or arcs, it may turn out to be impossible (or at least impossible with ruler-and-compass methods). In his comments to this proposition, Heath mentioned alternative proofs that were suggested by later authors and that bypass this problem (Euclid 1956, vol. 2, 146; Mueller 1981, 122).

A further interesting point to notice in relation with these six propositions of Book V is that their diagrams are all half way between the geometric (typical of proofs in Book II), and the purely arithmetic (such as in Book VII). Thus, on the one hand, the magnitudes are represented in the diagrams by lines, while they are meant as magnitudes of any kind, including areas, angles or volumes (but not numbers). On the other hand, these lines are not parts of constructions (like in the geometric proofs) but are just indicative of the magnitudes involved. As in the arithmetic type of proofs, the only way in which they are manipulated upon is that they are added to, or subtracted from, each other (Mueller 1981, 22). Again, this will appear as an important point below, because, as it is well known, Book VII develops a theory of proportions for natural numbers, which runs parallel to that of Book V, and in some medieval texts we find discussions about the need to follow Euclid in maintaining this separation, or the possibility to bypass it.

2.3 Book VII

Also in Book VII of the *Elements* we find propositions that involve distributivity-like properties. They focus on the question of "being a part" or of "being parts" of a given number and, like in Book V, they explore the issue equimultiplicity. There are two pairs of propositions, VII.5-VII.6 and VII.7-VII.8, that appear as parallel, respectively, to V.1 and V.5. They are formulated as follows:

VII. 5: If a number be a part of a number, and another be the same part of another, the sum will also be the same part of the sum that the one is of the one.

VII. 6: If a number be parts of a number, and another be the same parts of another, the sum will also be the same parts of the sum that the one is of the one.

VII. 7: If a number be that part of a number, which a number subtracted is of a number subtracted, the remainder will also be the same part of the remainder that the whole is of the whole.

VII. 8: If a number be the same parts of a number that a number subtracted is of a number subtracted, the remainder will also be the same parts of the remainder that the whole is of the whole.

Also in this case, the symbolical rendering suggested by Heath stresses very much a possible retrospective interpretation of these propositions as statements of distributivity of multiplication of numbers over their addition:[2]

VII.5: $\frac{1}{n}a + \frac{1}{n}b = \frac{1}{n}(a+b)$.

VII.6: $\frac{m}{n}a + \frac{m}{n}b = \frac{m}{n}(a+b)$.

VII.7: $\frac{1}{n}a - \frac{1}{n}b = \frac{1}{n}(a-b)$.

VII.8: $\frac{m}{n}a - \frac{m}{n}b = \frac{m}{n}(a-b)$.

But this rendering raises, beyond the historiographical issues already mentioned for any symbolic rendering of this kind, the additional problem that in Euclid's arithmetic there is nothing like the fraction $\frac{1}{n}$. Perhaps a closer symbolic approximation—not without its own limitations as a faithful rendering of Euclid—could be the following:

---

[2] A similar, though not identical, rendering appears in (Itard 1961, 90–97). (Taisbak 1971, 40 ff.) gives a completely different kind of symbolic rendering, conceived with the specific purpose in mind of avoiding any possible historical inaccuracy incurred by the use of modern algebraic symbolism. It is well beyond the intended scope of the present article to follow any kind of symbolic approach close to that of Taisbak. Still, it is interesting that, following his own point of view, Taisbak states explicitly (p. 43) that VII.5 and VII.6 "can be interpreted as the *Distributive Law*".

Fig. 2.9 Euclid's
Elements—VII.5

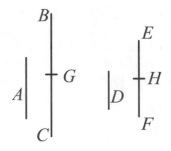

VII.5′: If $a = n \cdot b$ and $c = n \cdot d$, then $a + c = n \cdot (b + d)$.
VII.6′: If $m \cdot a = n \cdot b$ and $m \cdot c = n \cdot d$, then $m \cdot (a + c) = n \cdot (b + d)$.
VII.7′: If $a = n \cdot b$ and $c = n \cdot d$, then $a - c = n \cdot (b - d)$.
VII.8′: If $m \cdot a = n \cdot b$ and $m \cdot c = n \cdot d$, then $m \cdot (a - c) = n \cdot (b - d)$.

In mathematical terms this latter rendering is equivalent to Heath's, but it expresses more closely the spirit of the original. For one thing, contrary to what is the case with Heath's rendering, this one retains the "if … then" format of Euclid's original formulation. For another thing, in Heath's rendering, the four statements can be retrospectively seen as just different instances of the same general rule of distributivity. This raises the question of why Euclid would have chosen to state and prove them as separate results. Obviously, the straightforward equivalence between the four becomes possible only when written retrospectively in this way. This misleading interpretation is not fully avoided in the alternative rendering I just suggested, but it is at least somehow mitigated. I think it does provide a justification for seeing them as part of a family of related, distributive-like properties, without thereby having to assume the existence of a clear-cut, general idea of distributivity.

Euclid's four proofs are quite similar to each other and they are based on a rather straightforward counting of units. We can see the main argument through the example of VII.5, which is accompanied by the diagram in Fig. 2.9.

Here A is a part of BC, and D is the same part of another EF as A is of BC. The proposition then states that the sum of A, D is also the same part of the sum BC, EF as A is of BC. The core of the argument recapitulates that of V.1.[3]

Since we shall need this below, it is pertinent to say also a word about the proof of VII.6, which is accompanied by the diagram in Fig. 2.10.

The steps of the proof are the following:

(c.1) AC, GB represent parts of C that taken together make the number AB. Likewise, DH, HE represent parts of F that taken together make the number DE;

(c.2)   As many parts of C as taken together make AB that many parts of F taken together make DE;

---

[3] (Taisbak 1971, 41) suggests that the details of this proof indicate that Euclid implicitly takes for granted the associative and commutative laws for the addition of numbers, "implied as they are in his definitions of number".

Fig. 2.10 Euclid's
Elements—VII.6

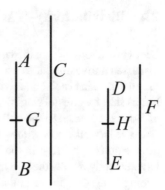

(c.3)   Hence "whatever part *AG* is of *C*, the same part also is the sum of *AG*, *DH*
        of the sum of *C*, *F*." [because of VII.5]
(c.4)   The same applies for *GB* vis-à-vis *C*, and for the sum *GB*, *HE* vis-à-vis the
        sum of *C,F*.
(c.5)   Finally: "whatever parts *AB* is of *C*, the same parts also is the sum of *AB*, *DE*
        of the sum of *C*, *F*."

                                                                            QED.

Notice that for lack of a more flexible language, Euclid took *two* parts in the proof
to mean "an arbitrary number of parts." In the argument, he referred to each of these
parts separately, *AG* and *GB*, and then moved to the conclusion for the sum. But he
did not say explicitly that this conclusion requires that the procedure can indeed be
validly applied to the case in which three, four, or any number of parts of *C* other
than two is added. It is quite clear, nevertheless, that the generality of the argument,
in this regard, is not compromised. Notice also that it is not known how many of
parts such as *AG* taken together make *C*, but we do know that as many as they are, *F*
is made of as many parts, each being like *DH*. Obviously then, the generality of this
part of the argument is not compromised in any sense, because we do not even need
to know into how many parts *C* or *F* are divided.

Referring to the symbolic rendering in **VII.6**, $\frac{m}{n} a + \frac{m}{n} b = \frac{m}{n}(a + b)$, we can say
that Euclid's argument is clearly valid for any value of *n*, because *n* is actually absent
in the argument. At the same time, the argument is presented in detail only for $m =$
2, and it is left implicit that it can be generalized to any value of *m*, though it is not
obvious how this could be done in the language typically available to Euclid.[4]

The proofs of VII.7–VII.8 amount to not much than similar counting of units as in
the previous two propositions, and both of them reduce their argument to a situation
where VII.5 can be directly applied.

---

[4] See (Itard 1961, 93–97) for what he considers to be a problematic aspect of Euclid's proofs for
VII.6 and VII.8. See (Taisbak 1971, 42–48) for additional, but quite different kinds of comments.

2.4 Distributivity-Like Results at Work

Let us take a look now at the ways in which the various distributivity-like propositions discussed above are put to use in the *Elements*. Taken together, it is remarkable that they are used in crucial places for proving many elaborate propositions in the treatise. Let us start by considering, with the help of Fig. 2.11, how the first ten propositions of Book II are used in the entire treatise.

As it is well known, and as shown in the diagram, a main application of the propositions appearing in the first part of the book is in the proofs of II.11,14. The latter embody results that are mathematically significant in themselves (respectively: cutting a segment in mean and extreme ration and constructing a square equal to a given rectilinear figure). These two also have important applications in later parts of the *Elements* (such as in IV.10, which is needed for constructing the regular pentagon, and which relies on II.11). But from the diagram, one also readily notices that Book X appears as a main focus of application of these ten results, and more specifically so, of II.4–II.9. Taken together with the applications in proofs in Books III, IV, XII XIII, we can say that these distributivity-like properties of area-formation do play an important role in the general economy of the treatise. A case of special interest for our discussion here is that of IX.15, in which proof, *arithmetic* versions of II.3,4 are used. I shall further comment on this issue right below.

As already pointed out, Euclid's basic approach in developing the first ten results of Book II was to prove each proposition on the basis of results of Book I alone. Technically speaking, however, it would be enough to prove II.1, and then all the rest could be proved by relying on this one proposition alone and without further recourse to Book I. This is, as we shall see below, the approach that Heron followed in his version of Book II. One can only speculate about the reasons behind Euclid's specific choice. Together with the issue of "geometric algebra" and the application of the propositions of Book II to prove additional propositions in the treatise, historians have discussed the possible significance of the approach followed by Euclid in his

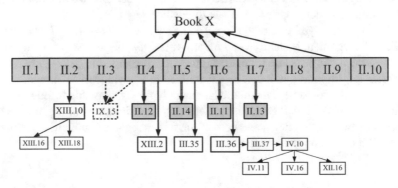

Fig. 2.11 Deductive dependence of distributive-like results in Euclid's *Elements*

proofs. Such discussions concern an assessment of the role of Book II as a whole—
and of some of its individual propositions separately—within the general economy
of the *Elements*. Ian Mueller, for instance, found that the evidence was inconclusive
to allow for such an assessment (Mueller 1981, 301–2):

> If one accepts II,11–14 as a goal of book II, one has an explanation for the presence of
> II,4–7, which are used in their proofs. ... As far as I am able to determine, there is nothing
> in the *Elements* themselves which makes the algebraic interpretation of these propositions
> more natural than the straightforward geometric one. On the other hand, the minimal use
> of II,1–3, 8–10, together with the generally loose connection between book II and books
> X and XIII, makes it difficult to feel confident about book II. ... What unites book II is
> the methods employed: the addition and subtraction of rectangles and squares to prove
> equalities and the construction of rectilinear areas satisfying given conditions. 1–3 and 8–10
> are also applications of these methods; but why Euclid should choose to prove exactly those
> propositions does not seem to be explicable.

Christian Marinus Taisbak, in turn, speculated about a possible explanation that
builds on some pre-Euclidean, ancient arithmetic traditions. In those traditions,
propositions similar to II.5, II.6, II.10 and II.11 appear in a natural way. Taisbak
added that II.1 should be considered, under this view, to have been added by Euclid
himself as a generalization of other propositions presented in Book II. He regretted,
moreover, that Euclid "did not tell us explicitly what is the meaning of it all, particularly what II.5 and II.6 are good for, although such knowledge is presupposed in
Book X from prop. 17 onward" (Taisbak 1993, 30).

Now, one may wonder to what extent, if at all, the medieval authors to be discussed
below ever asked themselves *historical* questions of this kind or what was their
view on issues related to these ones. We do know, as will be seen below, that they
occasionally came up with their own alternative *mathematical* approach to the proofs
and to the connection among the various propositions. But in order to complete the
picture of this part of the discussion, it is also important to stress that in several places
in the *Elements* we find proofs where distributivity-like properties of area formation
over addition are implicitly used without any explicit comment or further justification.
The typical case in point is in the proof of I.47, the Pythagorean Theorem, where
two rectangles with a common side are added to create a certain square. This can be
seen in the diagram of the proof, which is reproduced in Fig. 2.12.

A crucial step in the proof relies on an equality of areas:

$$\mathrm{Sq}(BC) = \mathrm{R}(BD, DL) + \mathrm{R}(CE, LE).$$

This step is mentioned in the proof by direct reference to the diagram, without
further ado. But on the other hand, this is precisely the kind of situation that is
handled, and duly proved, in II.1–3. How can we then explain that Euclid takes
the step in Book I for granted and without any comments, yet at the same time
finds it necessary to prove the same point in detail as a proposition in Book II?
This issue has been discussed by Ken Saito (2004, 164), who called attention to
the fact that the two rectangles added in the proof of I.47 are explicitly drawn (or
"visible") in the diagram. To the contrary, Saito indicates, rectangles appearing in

Fig. 2.12 Euclid's
Elements—I.47

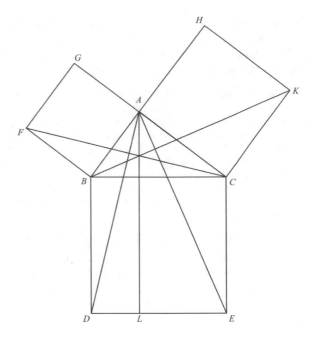

the proofs of II.1–3 do not arise as part of an explicit geometric construction in the corresponding diagrams (Saito calls them "invisible"). Justification such as provided by II.1 is necessary, Saito convincingly argues, because the rectangle to which the statement refers is invisible. Moreover, the crucial point of the proofs of II.1–3 is that the "invisible" rectangles referred to in the proposition are made "visible", so that this kind of "distributivity of area-formation", which is self-evident for "visible" rectangles, can be used.

Seen from this perspective, the first ten propositions of Book II are lemmas that are introduced beforehand, in preparation for later use in specific geometric situations. In those situations, it is possible to invoke the lemma without having to explicitly show in the diagram the relevant construction, which thus remains "invisible".

A most interesting example of this, which is relevant for our discussion here, concerns propositions II.12 and II.13. This example is not mentioned by Saito in his detailed account, but I think that it gives additional credence to his arguments, with the additional advantage that it is located within Book II itself, and put to use in additional propositions: not just II.1–3, as indicated by Saito, but also II.4 and II.7. The example is also interesting because this pair of propositions, II.12 and II.13, embody generalizations of the Pythagorean Theorem for triangles which are not right-angled. Moreover, their proofs, unlike that of I.47, are based, indeed, on "invisible" constructions such as identified by Saito and therefore their natural place is in Book II, rather than being added right after I.47 in Book I.

Let us start by formulating the propositions. The corresponding diagrams appear in Fig. 2.13.

II.12: In obtuse-angled triangles the square on the side subtending the obtuse angle is greater than the sum of the squares on the sides containing the obtuse angle by twice the rectangle

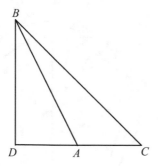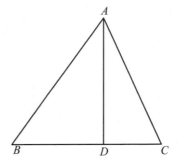

Fig. 2.13 Euclid's *Elements*—II.12, II.13

contained by one of the sides about the obtuse angle, namely that on which the perpendicular falls, and the straight line cut off outside by the perpendicular towards the obtuse angle.

II.13: In acute-angled triangles the square on the side subtending the acute angle is less than the sum of the squares on the sides containing the acute angle by twice the rectangle contained by one of the sides about the acute angle, namely that on which the perpendicular falls, and the straight line cut off within by the perpendicular towards the acute angle.

Notice that both propositions are reminiscent of a cosine-law kind of formula, in the sense that in each case there is a term that corrects the straightforward Pythagorean relation. We can write the relations as follows:

II.12: $Sq(CB) = Sq(AB) + Sq(CA) + 2 \cdot R(AC,AD)$.
II.13: $Sq(AC) = Sq(CB) + Sq(BA) - 2 \cdot R(CB,BD)$.

Let us see how II.4 is applied in the proof of II.12 (the role of II.7 in the proof of II.13 is similar):

(d.1) By II.4: $Sq(CD) = Sq(CA) + Sq(AD) + 2 \cdot R(CA,AD)$,
(d.2) Hence: $Sq(CD) + Sq(DB) = Sq(DB) + Sq(CA) + Sq(AD) + 2 \cdot R(CA,AD)$,
(d.3) By I.47: $Sq(CB) = Sq(CD) + Sq(BD)$,
(d.4) By I.47: $Sq(AB) = Sq(AD) + Sq(BD)$,
(d.5) Hence: $Sq(CB) = Sq(AB) + Sq(CA) + 2 \cdot R(CA,AD)$.

QED.

Below, I return to these two propositions.

Yet another illuminating example of this situation arises in XIII.10, which is the only place in the entire Euclidean treatise where II.2 is explicitly used. The proposition states than in an equilateral pentagon inscribed in a circle, the square on the side of the pentagon equals the sum of the squares on the sides of the hexagon and the decagon that are inscribed in the same circle (Fig. 2.14).

The proof is relatively complicated and we do not need to see all of its details here (but see (Taisbak 1993)). I just want to focus on the fact that one of the concluding steps of the proof requires that.

$$Sq\,(BA) = R(AB,BN) + R(AB,AN)\,.$$

Fig. 2.14 Euclid's
Elements—XIII.10

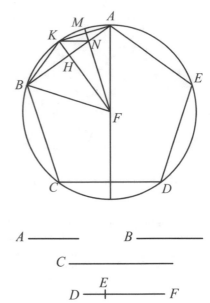

Fig. 2.15 Euclid's
Elements—IX.15

The step is justified on the basis of II.2, and thus the proof is completed without drawing any of these invisible figures explicitly. This also help explains why Euclid proved separately II.2-II.3, in spite of their appearance as no more than particular cases of II.1, the reason being that each of them needs to be ready for the specific situation in which it is going to be used (Saito 2004, 164).

Yet an additional focal point of interest that deserves special attention here is the proof of IX.15. Several stages of the argument rely on what can be seen as arithmetic versions of II.3 and II.4 (actually II.3 is not applied anywhere else in the *Elements*). We have seen that, because Euclid consistently adhered to the principle of separating realms throughout the treatise, separate versions are found of what can retrospectively be seen as similar distributivity-like results. Here, however, we have an interesting case of *transgressing* that principle.[5] The proposition in question is stated as follows:

> **IX.15**: If three numbers in continued proportion be the least of those which have the same ratio with them, any two whatever added together will be prime to the remaining number.

The accompanying figure is Fig. 2.15.

Here A, B, C are the three given numbers in continued proportion. Proposition VIII.2 warranties the existence of two numbers, DE, EF such that $A = DE^2$; $B = DE\cdot EF$; and $C = EF^2$. In addition, by VIII.22, DE, EF are mutually prime. Now, along the proof, Euclid considers several *products* involving DF, DE, EF and their squares. Using various propositions of Book VII (22, 24, 25), he can

---

[5] See (Mueller 1981, 180 ff.) for a broader discussion of Euclid's use of geometric arguments and analogies in arithmetical contexts. For example, Mueller indicates (p. 108) that in Book X, Euclid proves two *lemmatas* while invoking an arithmetic analogue of II.6.

establish certain relations of mutual primality among them. But in other places he also needs to consider cases where the numbers and their squares are *added* to one another, and it so happens that in Books VII and VIII, there is only *one* proposition (VII.28) that handles cases of adding mutually prime numbers. Euclid indeed invokes this proposition in order to prove that *DF* is relatively prime to both *DE* and *EF*. But in the other stages in the proof he deals with additions not covered for VII.28, and that can be interpreted as arithmetic cases of the situation covered (for the geometric case) by II.3 and II.4. Thus, for instance:

- The product of *FD, DE* is the square on *DE* together with the product of *DE, EF*.
- The squares on *DE, EF* together with twice the product of *DE, EF* are equal to the square on *DF*.

Now, in preparation for this proof, Euclid could have conceivably formulated and proved the purely arithmetic propositions needed here, but for some reason he declined to do so and preferred to deviate from his self-imposed, strict separation of domains. Again, we can only speculate about the reason for this choice, but it is interesting to mention that in the medieval texts discussed below we shall encounter an interesting twist to this peculiar situation. Al-Nayrīzī in the tenth century, and then Campanus in the thirteenth century, formulated a generalized version of IX.15 and added to it original "commentaries" most of which happen to be arithmetic versions of results from Book II. This is one of the most interesting issues that I discuss in detail below.

Let us now consider the way in which the six propositions of Book V discussed above are used in the *Elements*. This is schematically represented in Fig. 2.16.

In the lower dotted square I have indicated most, but not all, of the results that can be derived indirectly from the six propositions in question. Still, the picture is quite clear and interesting. While V.1 and V.2 are used *crucially* in some of the later

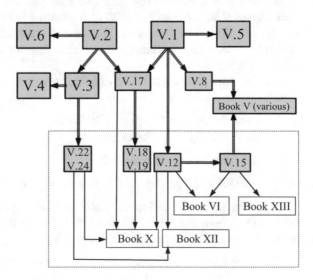

Fig. 2.16 Deductive dependence for distributive-like results in Euclid's *Elements*

proofs in the book, V.3 is used only in the proof of V.4, whereas V.5, V.6 are not used at all. Also, the derivatives of V.1 and V.2 do play important roles in Books VI, X, XII and XIII.

It is also pertinent to mention in this regard proposition V.24, which may be seen as expressing yet another kind of distributivity-like property of sorts. It reads as follows:

V.24: If a first magnitude have to a second the same ratio as a third has to a fourth, and also a fifth have to the second the same ratio as a sixth to the fourth, the first and fifth added together will have to the second the same ration as the third and sixth have to the fourth.

If we allow ourselves a symbolic rendering of this proposition then we obtain the following:

V.24: If $a{:}c :: d{:}f$, and $b{:}c :: e{:}f$, then $(a + b){:}c :: (d + e){:}f$.

It is important to notice that this proposition appears in the section of Book V, where Euclid dealt with "composition" of proportions. These are propositions such as V.17 or V.18, whose symbolic renderings are, respectively:

V.17: If $a{:}b :: c{:}d$, then $(a - b){:}b :: (d - e){:}f$.
V.18: If $a{:}b :: c{:}d$, then $(a + b){:}b :: (d + e){:}f$.

The proof of V.24 depends crucially on V.18, and it is noteworthy that in his well-known edition of the *Elements*, Robert Simson (1687–1768) indicated that Euclid's proof of V.24 can be easily modified to obtain, based on V.17, a result similar to that of V.24 but involving subtraction. Euclid never handled the case of subtraction, but as we shall see below, there is at least one medieval text (*Liber Mahameleth*) where this was actually done. What will turn out to be even more interesting, however, is that in that text V.24 is taken to be *the* source of justification for this distributive-like property of the product.

In the *Elements* there is only one other proposition whose proof relies on the main idea of V.24, but it does so with an interesting twist. This is proposition VI.31, which generalizes the Pythagorean Theorem by constructing on the sides of a right-angled triangle, not squares but rather any three figures that are "similar and similarly described" (Fig. 2.17).

In the proof, Euclid refers to non-specified geometric figures, and not necessarily to rectangles as in the figure. He shows separately that:

(e.1) $CB{:}BD ::$ (fig. on CB):(fig. on BA), and
(e.2) $BC{:}CD ::$ (fig. on BC):(fig. on CA),

and from here he deduces that:

(e.3) $CB{:}(BD + CD) ::$ (fig. on CB):(sum of figs. on BA and AC).

Completing this latter deduction requires something *close* to V.24, but not exactly the way the proposition is stated. Indeed, in V.24, it is the antecedents that are added in both ratios, and not as here in the proof of VII.31.

Finally, we can take a look at the way in which the six propositions of Book VII discussed above are used in the *Elements*. This is schematically represented in Fig. 2.18.

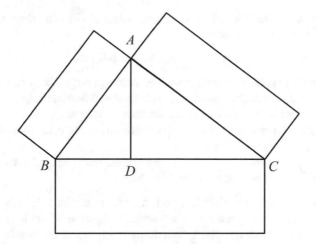

Fig. 2.17 Euclid's *Elements*—VI.31

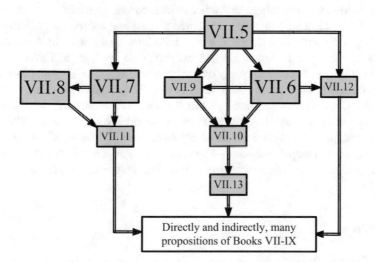

Fig. 2.18 Deductive dependence for distributive-like results in Euclid's *Elements*

In this case we see very clearly that VII.5 (both directly and indirectly via other propositions) provides an important tool that is consistently used for proofs throughout the three arithmetic books of the *Elements*. One important case to mention is that of VII.15,16, which embody the commutativity of the product (and are themselves also used, of course, in the proofs of many other propositions). A second important case is that of VII.11–13, three propositions needed to prove X.35. This latter proposition—in conjunction with several other propositions of Book VII that derive from VII.1—allows Euclid proving IX.36. This is the famous proposition that in modern terms can be understood as stating that if $2^p - 1$ is a prime number, then $(2^p - 1)\, 2^{p-1}$ is a perfect number.

The above survey may be now summarized as follows:

- Propositions involving distributivity-like properties appear in the *Elements* in three different realms (magnitudes, proportions and arithmetic),
- in each of these realms various cases are treated separately,
- different underlying assumptions, both explicit and implicit, are used in the various proofs, and in the various realms,
- distributivity-like results obtained in the separate realms are put to use in different ways in the overall economy of the *Elements*.

Mathematicians of later historical periods, whose works I discuss below, read the *Elements* from a perspective that departed from some of the basic underlying Greek classical conceptions. These conceptions concerned the nature of numbers and continuous magnitudes, the relationship among these two different kinds of mathematical entities, and the way they were put to use in various mathematical situations. Among other things, the rather strict separation between the three realms characteristic of Euclid's own approach, was revised and approached differently by the various later authors. Moreover, when these later mathematicians worked with positive numbers that included, beyond integers, also fractions and even irrational numbers, they could not directly rely on the canonical versions of those propositions that Euclid had proved in his arithmetical books. Book II above all, and to a lesser extent also Book V, provided many results that medieval mathematicians wanted to use in the arithmetic realm, which for them was broader than for Euclid. Their more flexible conceptions about the nature of numbers directly influenced, as we will see now, the ways in which distributivity-like results were used and justified, while some of the Euclidean propositions were themselves modified accordingly.

References

Acerbi, Fabio. 2003. Drowning by multiples. Remarks on the Fifth Book of Euclid's *Elements*, with special emphasis on Prop. 8. *Archive for History of Exact Sciences* 57 (3): 175–242. https://doi.org/10.1007/s00407-002-0061-y.

Euclid. 1956. *The thirteen books of Euclid's Elements*. Edited by Thomas Heath. 2nd ed., rev. with additions. 3 vols. New York: Dover.

Høyrup, Jens. 2017. What is 'Geometric Algebra', and what has it been in historiography? *AIMS Mathematics* 2 (1): 128–160.

Itard, Jean. 1961. *Les livres arithmétiques d'Euclide*. Paris: Hermann.

Mueller, Ian. 1981. *Philosophy of mathematics and deductive structure in Euclid's Elements*. Mineola, N.Y: Dover Publications Inc.

Saito, Ken. 2004. "Book II of Euclid's *Elements* in the light of the theory of conic sections." In *Classics in the history of Greek mathematics*, edited by Jean Christianidis, 139–68. Springer Netherlands. http://link.springer.com/chapter/. https://doi.org/10.1007/978-1-4020-2640-9_7.

Saito, Ken, and Nathan Sidoli. 2012. Diagrams and arguments in ancient greek mathematics: Lessons drawn from comparisons of the manuscript diagrams with those in modern critical editions. In

The history of mathematical proof in ancient traditions, ed. Karine Chemla, 135–162. Cambridge: Cambridge University Press.

Schneider, Martina R. 2016. "Contextualizing Unguru's 1975 attack on the historiography of ancient Greek mathematics." In *Historiography of mathematics in the 19th and 20th centuries*, edited by Volker R. Remmert, Martina R. Schneider, and Henrik Kragh Sørensen, 245–67. Trends in the history of science. Cham: Springer International Publishing. https://doi.org/10.1007/978-3-319-39649-1_12.

Taisbak, Christian Marinus. 1971. *Division and logos: A theory of equivalent couples and sets of integers, propounded by Euclid in the arithmetical books of the Elements*. Odense: Univ Pr of Southern Denmark.

Taisbak. 1993. A tale of half sums and differences ancient tricks with numbers. *Centaurus* 36 (1): 22–32. https://doi.org/10.1111/j.1600-0498.1993.tb00707.x.

Unguru, Sabetai. 1975. On the need to rewrite the history of Greek mathematics. *Archive for History of Exact Sciences* 15 (1): 67–114. https://doi.org/10.1007/BF00327233.

Chapter 3
Late Antiquity and Islamicate Mathematics

Abstract This chapter discusses some important instances of distributivity-like ideas appearing in the Islamicate tradition. It starts by examining al-Nayrīzī's additions to the *Elements* that appear in the framework of his report on Heron's *Commentary*. It also discuss briefly Abū Kāmil's version of propositions from Book II, as well as an interesting appearance of distributivity-like considerations surfacing in the work of al-Khayyām.

Keywords Distributive-like rules · The euclidean tradition · Islamicate mathematics · Arabic algebra · Heron · al-Nayrīzī · Abū Kāmil · al-Khayyām

In this section I discuss some important instances of distributivity-like ideas appearing in the Islamicate tradition. I start by examining al-Nayrīzī's additions to the *Elements* that appear in the framework of his report on Heron's *Commentary*. Al-Nayrīzī devoted a more focused attention to distributivity-like results than Euclid had done. I also discuss briefly Abū Kāmil's version of propositions from Book II, as well as an interesting appearance of distributivity-like considerations surfacing in the work of al-Khayyām.

3.1 Heron and Al-Nayrīzī

Heron of Alexandria wrote a *Commentary* of the *Elements* at the end of the first century AD In Book II, Heron preserved the purely geometrical spirit of Euclid's presentation, but some of his proofs are quite different from those attributed to Euclid. He followed a geometric approach that can be dubbed "operational", as opposed to Euclid's which, while geometric as well, is more "constructive" in spirit (Corry 2013, 652–54). In spite of this difference, however, Heron also maintained a clear separation between his handling of discrete and continuous magnitudes, along the separation that Euclid had instituted throughout the various parts of the *Elements*.

© The Author(s), under exclusive license to Springer Nature Switzerland AG 2021   27
L. Corry, *Distributivity-like Results in the Medieval Traditions of Euclid's Elements*,
SpringerBriefs in History of Science and Technology,
https://doi.org/10.1007/978-3-030-79679-2_3

Heron's ideas are known to us via al-Nayrīzī's commentary to the *Elements*, dating from the early tenth century AD. This is one of the earliest such commentaries written in Arabic. In Book II, alongside with reporting on Heron's ideas, al-Nayrīzī also added numerical examples of his own that were meant to illustrate the first five propositions. The inclination to do so may perhaps be explained against the background of the more flexible conception and use of numbers typical of the Islamicate tradition of which al-Nayrīzī was part. But it is important to stress that numerical interpretations of Book II could have hardly been accommodated within the specific approach followed in Euclid's proofs that were, as I said, "constructive". To the contrary, they found a more natural place within Heron's because, while geometrical, they were "operational" (Corry 2013, 661–62).

But al-Nayrīzī's own contribution went much further than just illustrating the propositions of Book II with numerical examples: he also incorporated into the arithmetical books of the *Elements* fully arithmetic versions of some of the propositions of Book II. He did so by adding to Book IX a section with his own comments. Also this move can be understood against the background of a more flexible conception of number that evolved in Islamicate mathematics, and along a less rigid separation between discrete and continuous magnitudes. Less rigid, it must be stressed, but not altogether inexistent. This move, at any rate, had a direct, visible influence on later medieval treatises, and specifically on Jordanus and Campanus as we will see below. Some of the details of Heron's and al-Nayrīzī's proofs are worthy of further examination here as they involve the reliance on distributive-like principles.

Heron asserted that II.1 is the only one among the fourteen propositions that "cannot be proved without drawing a total of two lines". For the remaining thirteen propositions, he stated that "it is possible that they be demonstrated with the drawing of one sole line" (Curtze 1899, 89).[1] There is no report on Heron's argument for II.1, and we may assume that it added nothing to Euclid's original. Al-Nayrīzī's numerical example for II.1 (p. 88), in turn, is embodied in Fig. 3.1.[2]

Al-Nayrīzī also reported on Heron's proofs for the other propositions in Book II. The most important feature of these proofs is that, unlike Euclid's original proofs, they do not rely directly on results of Book I. Rather, they rely first on II.1, and then on those other propositions from the same Book II that Heron gradually proved as he went. Thus, for instance, Propositions II.2,3 appear here as particular cases of II.1. Then, II.4 appears as directly derivable from II.1, relying also on the other two. We can gain some insight into Heron's approach by looking at his proof of II.4.

As already indicated, Euclid's proof of II.4 is based on I.43, which establishes the equality of the two rectangles *AG*, *GE* (see above, Fig. 2.4). Heron's proof is quite different (p. 92). For one thing, he did not even draw the full square of Euclid's original diagram, but rather, as here in Fig. 3.2, only the line *ba*, cut at an arbitrary point *g*.

---

[1] Unless otherwise stated, translations from Latin, Hebrew, French or German are mine.

[2] In his commentaries to the text of Al-Nayrīzī, Curtze (1899) added algebraic renderings to each proposition.

Fig. 3.1 Al-Nayrīzī's
diagram for Euclid's
Elements—II.1

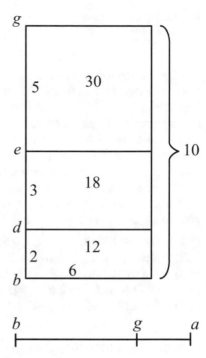

Fig. 3.2 Heron's diagram
for Euclid's *Elements*—II.4

The proposition states that the square on *ab* equals the sum of the two squares, one on *ag* and one on *gb*, together with twice the area contained by the lines *ag*, *gb*. Heron did not explicitly show any construction, but it is evident that he conceived this proposition, as well as the others in the book, as expressing properties of geometric figures. This same spirit is clearly reflected in the proof itself. His argument for II.4 can be schematically rendered as follows:

(f.1)   By II.2: Sq(ab) = R(ab,ag) + R(ab,bg),
(f.2)   But by II.3: R(ab,ag) = R(ag,bg) + Sq(ag),
(f.3)   Also by II.3: R(ab,bg) = R(ag,bg) + Sq(bg),
(f.4)   Hence: Sq(ab) = Sq(ag) + Sq(bg) + 2·R(ag,bg).

QED

Al-Nayrīzī's gives a numerical example, as in Fig. 3.3.

In addition, Al-Nayrīzī also gave the details of the calculation, namely, that the square on the entire length, 100, equals the sum 49 + 9 + 2·3·7.

$$a \qquad 7 \qquad g \quad 3 \quad b$$

Fig. 3.3 Al-Nayrīzī's diagram for Euclid's *Elements*—II.4

Fig. 3.4 Heron's diagram
for Euclid's *Elements*—II.7

Likewise interesting is the case of II.7, which Euclid proved in a way similar to
II.4, while relying directly on I.43, rather than on previous results of Book II. Also
here Heron's proof is quite different, and to be sure, it is decidedly "operational" in
the approach to the geometrical reasoning that it follows. His proof of II.7 (p. 92)
relies directly on II.4, as well as on II.3, and like in II.4, Heron did not draw the
full square of Euclid's diagram for II.7 (see above Fig. 2.5), or the other lines of the
construction. It focuses again on a line *ba,* cut at an arbitrary point *g* (Fig. 3.4).

The proposition states that the square on *ab* taken together with the square on *bg*
equals the square on *ag* taken together with twice the area contained by the lines *ab,*
gb. The argument of the proof for II.7 can be schematically rendered as follows:

(g.1)   By II.4: $Sq(ab) = Sq(ag) + Sq(bg) + 2 \cdot R(ag,bg)$,
(g.2)   Hence: $Sq(ab) + Sq(bg) = Sq(ag) + 2 \cdot Sq(bg) + 2 \cdot R(ag,bg)$,
(g.3)   But by II.3: $R(ab,bg) = R(ag,bg) + Sq(bg)$,
(g.4)   Or: $2 \cdot R(ab,bg) = 2 \cdot R(ag,bg) + 2 \cdot Sq(bg)$.
(g.5)   From (d.2), (d.4): $Sq(ab) + Sq(bg) = Sq(ag) + 2 \cdot R(ab,bg)$.

                                                                                            QED

In this case, al-Nayrīzī did not provide a numerical example.

But as already mentioned, al-Nayrīzī's own contribution went much further than
just illustrating the propositions of Book II with numerical examples: he took the
further step of incorporating into the arithmetical books of the *Elements* arithmetic
versions of propositions II.1–II.4. These arithmetic versions appear as commentaries
added to a result in Book IX (AN-IX.16) that, remarkably enough, does *not* appear
in Euclid's original version of the arithmetical sections of the *Elements*.[3] The details
of his proofs to these propositions deserve close attention.

The first interesting point to notice in these proofs is that their style is that typical
of the proofs appearing in Euclid's arithmetical books. In particular, the lines in
the diagrams indicate the numbers involved in the proof, but they are not used to
produce any relevant *geometric* constructions. Of particular importance is the fact
that multiplication is never represented here as area formation.

Al-Nayrīzī's diagram for his version of II.1, as it appears in the corresponding
commentary to AN-IX.16, is as in Fig. 3.5.

The line *hz* represents the product of *ab* by *gd*, whereas *kl* represents the product
of *ab* by *ge* and *ml* that of *ab* by *ed*. The proposition states that *hz* equals *km*. The
proof proceeds simply by spelling out each multiplication as the number of units by
which a number measures another:

---

[3] As a matter of fact, AN-IX.16 is a generalization of Euclid's IX.15. In Euclid's IX.15 three numbers
are added, whereas al-Nayrīzī adds an arbitrary number of numbers ("*Si fuerint numeri quotlibet
continue proportionales in sua proportione minimi,* …"). Curtze cites the proposition in a footnote
(p. 204), without in any way mentioning the discrepancy with the Euclidean original. As we shall
see below, Campanus followed al-Nayrīzī's formulation.

Fig. 3.5 Al-Nayrīzī's
diagram for the arithmetic
version of Euclid's
Elements—II.1

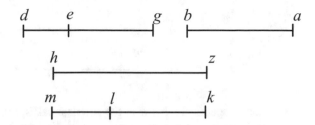

gd measures (*numerat*) *hz* by as many units as there are in *ab* whereas *ge* measures *kl* by
as many units as there are in *ab* and *ed* measures *ml* by as many units as there are in *ab*. ...
[Hence] the addition (*conjunctio*) *gd* measures *km* by as many units as there are in *ab*, and
hence the number *km* equals the number *hz*.

Translating back into "multiplication", al-Nayrīzī concludes that the area that is
obtained from *ab* and *gd* equals the addition of the two areas that are obtained from
ab and *ge* and *ab* and *ed*. And this is what we wanted to prove.

Thus we see that, while attempting to incorporate these results into the corpus
of arithmetical knowledge displayed in the *Elements*, al-Nayrīzī nevertheless abode
by the basic separation of realms. He did not import into the arithmetic books the
kind of geometric reasoning with *continuous* magnitudes used by Euclid in Book II,
but rather developed a proof that followed Euclid's own constraints for dealing with
discrete quantities. Recall my explanation above in relation to V.1-V.6, concerning
the conception of addition not as a binary operator, but rather as a gathering together
of multitudes of instances of a given magnitude. Al-Nayrīzī deals here with discrete
magnitudes and his proof is based on implicitly rearranging, according to the need,
the instances of the magnitudes that appear in the said multitudes. As we will see,
such rearrangement are performed explicitly in the work of Campanus as the basis
of some of his arguments.

Al-Nayrīzī also formulated arithmetical equivalents of II.2,3, and it interesting to
look at the proofs. We just saw that in his version of Book II, al-Nayrīzī relied on
II.1 for proving these two propositions. This is what Heron did, and it is different
from what Euclid had done (he did not rely on II.1 for proving II.2,3). Still, both
Heron and Euclid followed geometric arguments, though different from each other.
Now here, in the arithmetic equivalents of the two propositions, II.2 and II.3, Heron
went like Euclid; that is, he did not rely on (the arithmetic version of) II.1, but rather
rehearsed Euclid's original argument for II.2,3 (though now in arithmetic version).
Likewise, he also proved the arithmetic version of II.4 by a repeated application of
II.2 (pp. 205–207). Thus, al-Nayrīzī clearly wanted to stress the autonomous, purely
arithmetic characters of these propositions as presented in his commentary to IX.16.

Likewise interesting for our discussion here are al-Nayrīzī's comments to propositions V.1,2. First, concerning V.1, he indicated a possible difficulty in the argument of the proof (pp. 169–170). In order to see what he had in mind, consider the
accompanying diagram (Fig. 3.6).

Recall that in the argument of the proof, *ba*, *dg* represent equimutiples of *e*,
z respectively, and it is required that the each of the latter be cut from each of

Fig. 3.6 Al-Nayrīzī's
diagram for Euclid's
Elements V.1

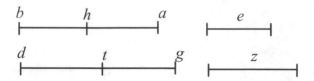

the former, respectively. Now, in the simplest cases, this raises no difficulties. For example, if the two given magnitudes *ab* and *gd* are lines, he wrote, then we can simply invoke Euclid's I.3 ("Given two unequal straight lines, to cut off from the greater a straight line equal to the less"). In the case where the magnitudes are arcs, al-Nayrīzī invoked Book III as providing the necessary justification. Most likely he had in mind a combination of propositions such as the following two:

> **III.27:** In equal circles angles standing on equal circumferences are equal to one another, whether they stand at the centres or at the circumferences.
>
> **III.34:** From a given circle to cut off a segment admitting an angle equal to a given rectilinear angle.

Also for the case when the magnitudes are arcs al-Nayrīzī declared that no problem arises, but he did not explain why. It is also plausible that his implicit justification for this case might have been related, via arcs of circles, to the same two results of Book III, in conjunction with the following one:

> **VI.33**: In equal circles angles have the same ratio as the circumferences on which they stand, whether they stand at the centres or at the circumferences.

But for the case when the magnitudes involved are bodies, al-Nayrīzī indicated that the necessary operation of subtraction becomes "impossible" ("… *tunc illud erit impossibile*."). Nevertheless, he asserted, the existence of multiples is assumed in this case, only in order to imagine that if the number of times that *e* measures *ab* is two, then the number of times that *z* measures *gd* is also two, or if it is half of this then also *z* is half *gd*, and so on for any multiplicity whatsoever.

Thus, al-Nayrīzī considered this proposition as embodying several different, but *specifically geometric* situations, each of which required its own kind of justification. In other words, he was not thinking of "magnitudes" as a completely general concept on which we can argue in abstract terms, without a specific justification for each case. Properties of equimultiplicity, so it seems, were for him differently rooted in basic properties specific to each kind of magnitude that can be considered.

His comments on V.2 are somewhat cryptic and we can at best conjecture what was that he had in mind. Al-Nayrīzī asserted that "there is nothing at all except the order of the branches of knowledge, of which the first is arithmetic, which is about numbers, and after which comes geometry. The proposition therefore demonstrates the basic, necessary principles which we will discover in this theory" (p. 170). In trying to understand what he had in mind here recall the original formulation of the proposition, which reads as follows (Fig. 3.7):

> **V.2**: If a first magnitude be the same multiple of a second that a third is of a fourth, and a fifth also be the same multiple of the second that a sixth is of the fourth, the sum of the first

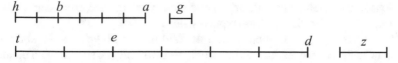

Fig. 3.7 Al-Nayrīzī's diagram for Euclid's *Elements*—V.2

and fifth will also be the same multiple of the second that the sum of the third and sixth is of the fourth.

The argument of his proof reads as follows:

Once we know that *g* measures *ab* according to the number of times that *z* measures *ed* then the times that *g* measures *ab* and that it measures *bh* equal the number of times that *z* measures *de* and *z* measures *et*. Hence, the enumeration of multiples *ah* equals the enumeration of multiples *dt*, and this is what we wanted to prove.

As we just saw, in his treatment of V.1 al-Nayrīzī found it relevant to speak about the meaning of the proposition with respect to various kinds of geometrical magnitudes. Now in discussing V.2 he invoked the importance of "the order of the branches of knowledge" and then limited himself to a general argument presumably valid for all kinds of magnitudes. Perhaps he meant to say that it is not necessary to discuss the various cases separately because his argument covers "the basic, necessary principles that we will discover in this theory". Admittedly, this conclusion is somewhat conjectural and hence inconclusive. Much less can we know how later readers interpreted it, or if they paid attention to this remark at all.

We thus see how al-Nayrīzī —in relation to a variety of Euclidean propositions where distributivity-like conditions are discussed—combined in original ways ideas that in the original conception of the *Elements* where treated separately, and added his own contribution. As with other noteworthy issues related to his commentary of Heron, the approach developed in this context turned out to have a significant influence, as will be seen below, on the important texts of Jordanus Nemorarius and Campanus de Novara.

3.2 Abū Kāmil: *Algebra*

The next case I want to consider within the Islamicate mathematical tradition concerns Abū Kāmil's treatise on algebra, where we find an arithmetical distributivity-like law of multiplication over addition that is grounded on essentially geometric considerations. This way to handle a basic property of an arithmetical operation reflects basic tension permeating the entire treatise, which, at the same time, is illustrative of broader issues arising in the Islamicate mathematical culture. Such issues derive from the attempt to combine and reconcile diverging conceptions and practices found at the sources of this culture. While the numbers handled in the arithmetical parts of the *Elements* were always positive integers, the practical traditions from which Islamicate arithmetic arose were at ease, from very early on, with fractions and irrational roots (typically positive).

Al-Khwārizmī's had codified, within the tradition of *al-ǧabr wa'l-muqābala*, a system of basic techniques for solving problems involving the square of an unknown quantity. The important corpus that arose from this codification is what we mean by Arabic algebra in what follows. Most importantly, he introduced new kinds of argumentation for justifying the validity of the techniques that had traditionally been used by his predecessors for solving problems of this kind. He remained close, on the one hand, to arguments previously found in the practical traditions, but, on the other hand, he followed a geometric style which was foreign to it (Oaks 2009). The resulting geometric style differed in important senses from the classical Euclidean one, being less rigorous and more intuitive or "visual" in its essence (Høyrup 1986; Oaks 2014, 44). This kind of cross-cultural and incomplete borrowing, from the Greek and from other traditions, has been associated also with the work of al-Uqlidisi (c. 920–c. 980) (Saidan 1978), Ibn al-Banna (1256–c. 1321) and al-Farisi (1267–1319) (Oaks 2019).

But it was Abū Kāmil (c. 850–c. 930) who introduced in a more systematic way the Euclidean style of proof into the arguments for justifying the techniques of Arabic algebra. Adopting Euclid's style and its concomitant basic conceptions on numbers and magnitudes, and trying to make it fit into the mold of practical Arabic arithmetic and algebra in which Abū Kāmil's work was rooted, implied some interesting challenges. For example, while the operations of multiplication, division, or root extraction are closed in the domain of numbers, in the realm of magnitudes they involve a change in dimension. But then, in the practical traditions from which Abū Kāmil's stemmed, the idea of geometric magnitudes as having numerical measure was—contrary to the spirit of the classical Euclidean view—a matter of routine. These underlying tension is evidently present in Abū Kāmil's approach and in his way of providing a Euclidean-style justification to arithmetical rules (Corry 2013, 655–61; Oaks 2011a, b). This is in particular the case when it comes to distributivity-like properties of the kind we are discussing here.

An example of particular interest for us appears in Abū Kāmil's discussion of rules for multiplying expressions involving the unknown or its square. Results and methodological approaches taken from the arithmetic parts of the *Elements* are interestingly combined here with results and proofs originally meant to deal with continuous magnitudes, such as such as those of Book II. Thus, for instance, the case "How much is ten *dirhams* and a thing by a thing?", or its parallel, "How much is ten *dirhams* less a thing by a thing?" Retrospectively seen, these can be thought of as a relevant instance of a distributivity-like situation, as reflected in the fact that, when symbolically rendered they correspond to $(10 + x) \cdot x = 10x + x^2$, or $(10 - x) \cdot x = 10x - x^2$.

Abū Kāmil's proofs of such cases consist, essentially, in representing the relevant arithmetic situation via a geometric diagram such as in Fig. 3.8 (Sesiano 1993, 342).

Here AB is taken to be 10 and BG to be the thing, while the rectangle AD is the product whose value we are looking for. The figure is meant to represent the fact that segments BE and GD are equal, while also GD and GB are equal. Hence BE equals

Fig. 3.8 Abū Kāmil's
diagram for solving a
problem with squares of the
unknown

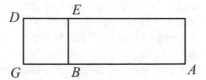

GB, which is the thing. Accordingly, then, *BD* is the square of the thing. Hence the rectangle *AD* is ten things and a square (of the thing), as stated in the rule.

Now, we saw above that in Euclid's proof of II.1 the distributivity of rectangle-formation for invisible figures derived from a geometric property (concatenation of rectangles) that is evident for visible figures. What we find here is a discussion of a rule of manipulation for the unknown quantity and its square (which are numbers). This arithmetic rule, however, is justified with the help of a geometrical argument (conceived to be used with continuous magnitudes). The important point is that it was not at all strictly necessary for Abū Kāmil to provide such a geometric justification. To be sure, in specific algebraic situations, he recurrently reduced the degree of an equation by dividing each term by the lowest power in a few problems. Thus, for instance: "a *māl māl* and twelve *māls* and a fourth of a *māl* equal nine cubes. Return everything you have to a *māl*, to get: a *māl* and twelve *dirhams* and a fourth of a *dirham* equal nine things" (Rashed 2012, 445). In modern algebraic terms this can be rendered as reducing

$$x^4 + 12^1/_4 x^2 = 9x^3 \text{ to } x^2 + 12^1/_4 = 9x.$$

In this case no specific justification was invoked, certainly not a geometric one, and yet in his treatise on algebra, Abū Kāmil did include proofs of the kind just discussed above, and he did so for the benefit of any of his Arabic readers trying to master the methods and the essence of the kind of geometrically-oriented algebraic thought that he was promoting. The justification he provided in this case, moreover, is embodied in a situation evidently similar to that of Euclid's II.1.

The Latin version of Abū Kāmil's treatise eventually became widely read among Europeans interested in learning algebraic techniques, and thus became acquainted with this way of handling distributivity-like properties related to arithmetical contexts. They learnt, that is, that such situations involving numbers are conveniently understood as situations that may or need to be supported by some kind of geometric justification. We shall find this approach repeated in many of the texts discussed below.

3.3 Omar Khayyam: *Algebra*

Yet another interesting perspective on distributive-like principles in the Islamicate tradition appears in the famous treatise of 'Umar al-Khayyām (1048–1131), *Algebra*, dating from the last third of the eleventh century. In this treatise, and to a lesser extent also in an earlier work, *On the division of a quadrant of a circle*, al-Khayyām followed a direction that is a mirror-image of the one just mentioned in the case of Abū Kāmil: he appropriated the practical methods of arithmetical problem-solving as a convenient tool for use in classical geometric problem-solving. In order to do so, however, he first needed to imbue those methods with a more rigorous perspective, firmly basing them on strict geometrical foundations (Oaks 2011a, b).

In al-Khayyām's work we find a combined handling of numerical situations while relying on two parallel traditions: on the one hand, the positive integers typical of Euclid and the Greek classical tradition, and on the other hand, the "continuous" numbers which Arabic algebraists adopted in their works even prior to al-Khayyām, and which included fractions and irrational roots essentially without limitations.

In what concerns coefficients, al-Khayyam worked with rational ones, as did many other medieval algebraists, but he did not often show them in the *Algebra*, because the cases treated in this treatise were designed to illustrate a general method. Irrational coefficients did not appear in any of the premodern traditions involving such algebraic methods (Oaks 2017). Still, he treated these numbers together with their powers under the more general, abstract idea of "quantity", which comprises also the continuous magnitudes of various dimensions. Under this perspective, the equations to be solved comprise "numbers", "things", "squares" (*māl* pl. *amwāl*), "cubes", "square-squares" (*māl māl*) (Corry 2015, 111–16).

The presence of these "square-squares" magnitudes (already mentioned above for the case of Abū Kāmil), should not lead us to believe, that al-Khayyām's algebraic methods involved a thoroughly abstract conception of magnitudes detached from geometric meanings. To be sure, the methods for which al-Khayyām's *Algebra* is best known are those that concern equations involving cubes, squares, roots and numbers. And in relation with "square-squares" he was very clear in stating, in both the *Algebra* and the *Quadrant*, that they do not have any meaning in continuous magnitudes, for "how could [a square] be multiplied by itself?" Algebraic methods (in the sense of the Islamicate tradition, of course, not the much later symbolic one), he stressed in the *Quadrant*, are related only to the other four genera. Algebra, to be sure, is part and parcel of geometry (Rashed and Vahabzadeh 2000, 171):

> Those who think that algebra is a method to determine unknown numbers think the unthinkable … Algebra is something geometrical which is demonstrated in Book II of the *Elements*, in Propositions 5 and 6 thereof.

While he emphasized this point again in *Algebra*, he turned there to the use of algebraic techniques for solving also problems involving numerical solutions, while providing detailed geometric proofs for each case considered.

When applying his algebraic methods to solve a geometric problem al-Khayyam designated some line as having a numerical length, like 10. He then gave algebraic names to other lines (and areas), like "a thing", and set up an equation involving them. Once the equation had been established from a geometry problem, there were no dimension to its terms, and hence all numbers were homogeneous. The product of two numbers, say 5 by 6, could be added back to 5, just as he could add "a thing" to "a cube" to get "a cube and a thing". But then, when the simplified equation in geometric terms was reinterpreted to construct the solution, the powers were again associated with magnitudes of a particular dimension. At this stage, each term in the equation was regarded as having the same dimension as the highest power term.

As the rules for solving simplified equations involve interpreting the multitudes of the powers arithmetically or geometrically, and thus they were required to be positive, al-Khayyām listed a total of twenty-five possible cases of equations involving cubes, squares, roots and numbers. In cases involving only squares and roots, and no cubes, al-Khayyām gave both arithmetic and geometric solutions. "When the subject of the problem (for irreducible cubic equations) is an absolute number—he wrote—it was not feasible either for us or for any one of those concerned with the art—and possibly someone else will come to know it after us..." (Rashed and Vahabzadeh 2000, 114) (and recall that, indeed, numerical solutions for equations involving also cubes became known only with the work of Cardano in the sixteenth century).

For instance, if the equation to be solved involves only squares (for example, the case: "squares and things equal numbers") the said "numbers" measure planes. Typically, but not always, they are seen as a rectangular surface one of whose sides is a "unit" while the second is a line equal in measure to the given number. In constructions related to cubic equations, "numbers" measure a solid. Typically, they may be a rectangular parallelepipedal solid whose base is the square of the unit and whose height is equal to the said number. While the dimension of "number" changes according to the degree of the equation under consideration, the other elements, "thing", "square", "cube", represent the same mathematical idea in all situations.

It is interesting to notice that the use of a "unit length" in geometry belongs in itself to another important thread within the broader story leading to the consolidation of an algebraic language for geometry. The thread started in late antiquity and received considerable attention in the Islamicate mathematical tradition, like here in the work of al-Khayyām. It would reach a peak, of course, in the seventeenth century with Descartes' *La Geométrie*, where it allowed the definition of a product of two segments yielding another segment, rather than a rectangle as with the classical geometry of the Greeks (Bos 2001, 293–301). Indeed, in the Islamicate context, it was routine to think of geometric magnitudes as having numerical measure. Al-Khayyam made this idea rigorous, by identifying the numbers of the algebraists with the measures of magnitudes. Since he assumed a unit line, these numbers were without dimension.

Now, half-way between the equations involving cubes and those involving only squares al- Khayyām also dealt with the so-called "reducible" equations, which involve cubes in their formulations, but which do not involve an independent term, and hence can be reduced to one of the cases of the lower degree involving only squares. What is interesting for us in the framework of the present account is the rather complicated way in which this apparently straightforward reduction is performed. In hindsight we can acknowledge the tacit reliance on a distributive-like properties of the magnitudes and operations involved. Let us see how.

There are three such reducible cases. Their basic formulations and those to which they can be reduced are the following:

A cube and squares equal things ↔ A square and things equal numbers,

A cube and things equal squares ↔ A square and numbers equal things,

A cube equals squares and things ↔ A square equals things and numbers.

The basic idea behind the reductions is rather self-evident in all three cases, and if we write them symbolically in modern terms, then it is even easier to see that they are all quite trivial:

$$x^3 + bx^2 = cx \leftrightarrow x^2 + bx = c,$$

$$x^3 + cx = bx^2 \leftrightarrow x^2 + cx = b,$$

$$x^3 = cx + bx^2 \leftrightarrow x^2 = cx + b.$$

This kind of straightforward reduction of the degree of an equation, as mentioned above in relation with Abū Kāmil, is also found in the work of al-Khayyām. Thus, for instance, in the *Quadrant* he reduces a fourth degree equation to a cubic equation by dividing everything by a "thing", without any proof or further comment, as described in the following passage (Rashed and Vahabzadeh 2000, 170):

> Two squares and one-tenth of a tenth of a squared-square [*māl māl*] are equal to twenty things and one-fifth of a cube. Then we divide the whole by the thing so that it be reduced to the lesser four genera which are in this ratio. The result of the division will then be: One-tenth of a tenth of a cube and two things are equal to one-fifth of a square and twenty in number.

In the context where this passage appears, the unknown quantity and its powers are not geometric magnitudes, but rather dimensionless entities. But the situation changes when we move to the three cases treated in his *Algebra*. Here it is the reduction itself that makes manifest that the geometric solutions to two equations are the same, while the division takes place in the realm of continuous numbers. Al-Khayyām devotes a full argument and a detailed geometrical construction in order to prove each of the cases, and the question arises why he followed this approach in this case. In order to try and make sense of it, we need to look at the proofs. Let us then consider one example in detail, namely the first case ("A cube and squares equal things" – Fig. 3.9):[4]

> Take a cube *ABCDE* and extend the straight line *AB* up to *Z*, so that *AZ* equals the number of squares. Complete the solid *AZHTCD* so that it is an extension of the cube *ABCDE*, as this is usually done. Then, the solid *AT* equals the number of squares, and the solid *BT*, which equals the cube together with the given number of squares, will be equal to the given number of things. Construct now a square *K* equal to the given number of things. The thing is the edge of the cube, that is, *AD*. Thus, the square *K*, when multiplied by *AD*, equals the

---

[4] I combine here from two available translations, (Woepcke 1851, 25–26; Linden 2016, 120–21), and at the same time slightly correct a detail. Woepke writes: "le rectangle *HA* … ést ègal à ce nombre de racines (de *CB*) qui avait été donné pour les carrés." I think that the addition "(de *CB*)" is misleading. Linden, on the other hand, speaks directly of the rectangle *K*, without indicating beforehand that it was introduced to satisfy a certain important property, as stated here.

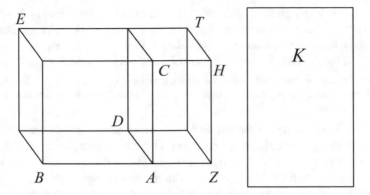

Fig. 3.9 Diagram accompanying the reduction of al-Khayyām's equation. "A cube and squares equal things."

given number of edges [i.e., of "things"]. On the other hand, rectangle *HB* multiplied by *AD* yields the cube plus the given number of squares. But these two solids are equal, namely, *BT* and the solid constructed on *K* with height *AD*. Accordingly, their bases are reciprocally proportional to their heights. But as their heights are equal so are their bases. But the base *HB* is equal to the square *CB* together with the rectangle *HA*, which is equal to the given number of things. Therefore *K*, which is the number of things (given by the squares) equals the square and the number of things given by the square.

QED

It is helpful to render this reasoning in a more schematic, symbolic formulation, in order to be able to comment on it. We can do that using the notation P(*DA*,*BC*) to indicate a parallelepipedal solid with height *DA* and base *BC*:

(h.1)   *ABCDE* = P(*DA*,*BC*) [this is the given cube, x^3],

(h.2)   Take *AZ* = *b* [*b* is the "coefficient" of the squares, and hence it is of dimension 1],

(h.3)   Construct *AZHTCD* = P(*DA*,*AH*) [i.e., bx^2 – jointly of dimension 3],

(h.4)   *BAZHTE* = P(DA,*BH*) [i.e., $x^3 + bx^2$].

(h.5)   P(*DA*,*BH*) = *cx* [this is what the equation states, with *c* of dimension 2];

(h.6)   Construct *K* = *c* [*K* of dimension 2],

(h.7)   Hence: P(DA,*K*) = *cx* [of dimension 3, of course],

(h.8)   But also by (h.4) P(DA,*BH*) = $x^3 + bx^2$, and hence P(DA,*K*) = P(DA,*BH*).

(h.9)   Hence *K* = *BH*, because they have the same height [by Euclid's XI.34],

(h.10) Hence *K* = *CB* + *HA*, but *HA* = *bx* [since *AZ* = *b*] and *CB* = x^2 [by construction],

(h.11) Hence, by (6) and (10): $x^2 + bx = c$.

QED

A main question that arises in relation with the proof concerns the role of the somewhat arbitrary introduction of a rectangle *K*, and why is it a necessary step in the proof. Let us try to address it. The geometrical versions of the entities involved in the original problem are three-dimensional rectangular solids, whereas those of the reduced one are the two-dimensional bases of those solids, and the aim is to show that a solution to the cubic equation is also a solution to the quadratic equation. The key tool applied by al-Khayyām for doing so is Euclid's XI.34, which reads as follows:

XI.34: In equal parallelepipedal solids the bases are reciprocally proportional to the heights; and those parallelepipedal solids in which the bases are reciprocally proportional to the heights are equal.

This is the crucial step of the proof and it appears here in (h.9). Constructing the geometric version of one side of the equation, $x^3 + bx^2$, is straightforward and this is done in steps (h.1-h.4). The other side requires additional steps. In (h.5) we see that $P(DA,\textbf{\textit{BH}}) = cx$, with DA representing x, which is not enough to conclude that BZH is indeed the unknown magnitude x. In order to apply X.34, he needs to construct a second parallelepipedal solid, equal to $BAZHTE$, and for this he introduces K, which because of X.34 ends up being equal to BH. Thus, rectangle K allows at the end of the proof to equate c with $x^2 + bx$, which was obtained by lowering the rank of $x^3 + bx^2$ by directly looking at the diagram.

The combined effect of the last few steps in the proof helps coming up, in a somewhat roundabout manner, with what a distributive-like result would provide in one simple step, which can be observed transparently in the symbolic representation of the equation, namely:

$$x^3 = cx + bx^2 \;\leftrightarrow\; x(x^2 + bx) \;\leftrightarrow\; x^2 = cx + b.$$

As already indicated, elsewhere in al-Khayyām's works, reductions leading from expressions such as in the left-hand side to those of the right-hand side are performed without any further comment, but here, where he wants to follow all the rigorous steps necessary for the reduction, the middle step (which embodies the distributivity which is crucial for completing the proof) is implicitly performed only *geometrically*, with the help of Euclid's proposition and of surface K which is needed in order to create the second parallelepipedal solid equal to the first one.

This situation would be further clarified by looking at the other two reductions discussed by al-Khayyām in this section of the *Algebra*. In consideration with what is already a very long discussion, I will refrain from doing so in detail. And yet I find it pertinent to look at Fig. 3.10, which represents the diagram accompanying the proof in the third case:

$$x^3 = cx + bx^2 \;\leftrightarrow\; x^2 = cx + b.$$

This is a cube built on the unknown magnitude, and it is partitioned into two solids, representing the addition $cx + bx^2$. Rather than relying on XI.34, al-Khayyām relies here on XI.32, which read as follows:

XI.32: Parallelepipedal solids which are of the same height are to one another as their bases.

Notice that this proposition *does not* require the existence of two equal solids, and indeed, nothing like the area K such as was necessary in the case previously

Fig. 3.10 Diagram
accompanying the reduction
of al-Khayyām's equation "A
cube equals squares and
things."

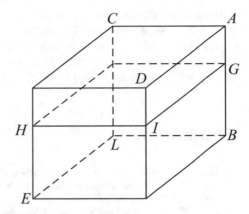

discussed (and also in the second case, not discussed here). The proposition states
that the ratio of the top solid is to the bottom one as the ratio of the base *GC* is to
GL, and this is what justifies the required reduction.

Al-Khayyām himself, to be sure, stressed the fundamental centrality of geometry
for what he achieved in this part of his treatise in terms of reducing the equations
involving cubes. He thus wrote (Woepcke 1851, 28):

> Inasmuch as these proofs [of the three kinds of equations] are understood other than in this
> way (i.e., geometrically; while initially they were conceived from a purely arithmetic point
> of view), the art of algebra does not become truly scientific, even when this method requires
> that we address some difficulties.

In his *Division of a quadrant*, Al-Khayyam had reduced degrees of equations
where possible, including a fourth-degree equation, without any further explanation
or justification, and yet in *Algebra* he found it necessary to provide rigorous proofs
for the same. These are not proofs for reduction of degree in general, but rather three
specific equation types, each worked out individually using a specific geometric
argument. His aim was to show that the geometric constructions of the solutions
to these cubic equations are the same as the constructions for the corresponding
quadratic equations that he had already given, and the crucial step that allowed him
to do in a tacit, but meaningful reliance on a distributivity-like property.

To be sure, al-Khayyām did not formulate any kind of explicit distributivity rules,
nor are such rules found in other arithmetic or algebraic contexts of Islamicate mathematics. Such rules represent special cases of the more general rule for multiplying
composite numbers as understood by Islamicate mathematicians and they can be
interpreted as a consequence of the very way in which the operation of multiplication was understood. In the case considered here, approaching an algebraic situation
from a rigorous point of view akin to the Euclidean tradition makes it more difficult
to simply reduce the dimensions of the quantities involved and hence the implicit
use step involving a distributive-like operation must be done with great care, while
relying on the adequate proposition taken form the *Elements* to justify it.

To conclude this section, it is important to stress that in spite of their originality and forcefulness, al-Khayyām's contributions to algebra had no noticeable impact on practical Islamicate algebra. Nor did it have, as a consequence, any influence on the development of algebraic thought in the Latin middle ages or on Hebrew sages writing on this topics (Oaks 2011a, b, 71–72).

References

Bos, Henk J. M. 2001. *Redefining geometrical exactness*. Sources and Studies in the History of Mathematics and Physical Sciences. New York, NY: Springer. http://link.springer.com/. https://doi.org/10.1007/978-1-4613-0087-8.

Corry, Leo. 2013. Geometry and arithmetic in the medieval traditions of Euclid's *Elements*: A view from Book II. *Archive for History of Exact Sciences* 67 (6): 637–705. https://doi.org/10.1007/s00407-013-0121-5.

Corry, Leo. 2015. *A brief history of numbers*. Oxford: Oxford University Press.

Curtze, Maximilian, ed. 1899. *Anaritii in decem libros priores elementorum Euclidis commentarii: Ex interpretatione gherardi cremonensis in codice cracoviensi 569 servata*. *Euclidis Opera Omnia*, I.L. Heiberg et H. Menge (eds.), Supplementum. Leipzig: Teubner.

Høyrup, Jens. 1986. Al-Khwârizmî, ibn Turk, and the liber Mensurationum: On the origins of Islamic algebra. *Erdem* 2: 445–484.

Linden, Sebastian. 2016. *Die algebra des Omar Chayyam*, Berlin: Springer.

Oaks, Jeffrey A. 2009. Polynomials and equations in Arabic algebra. *Archive for History of Exact Sciences* 63 (2): 169–203. https://doi.org/10.1007/s00407-008-0037-7.

Oaks, Jeffrey A. 2011a. Al-Khayyām's scientific revision of algebra. *Suhayl* 10 (1): 47–75.

Oaks, Jeffrey A. 2011b. Geometry and proof in Abū Kāmil's *Algebra*. In *Actes Du 10'eme Colloque Maghrébin sur l'histoire des mathématiques arabes (Tunis,29–30–31 Mai 2010)*, 234–56. Tunis: L'Association Tunisienne des Sciences Mathématiques.

Oaks, Jeffrey A. 2014. Abū Kāmil, algèbre et analyse diophantienne. Édition, traduction et commentaire by Roshdi Rashed. *Aestimatio* 11: 24–49.

Oaks, Jeffrey A. 2017. Irrational 'coefficients' in renaissance algebra. *Science in Context* 30 (2): 141–172. https://doi.org/10.1017/S0269889717000096.

Oaks, Jeffrey A. 2019. Proofs and algebra in Al-Fārisī's commentary. *Historia Mathematica, on Mathemata: Commenting on Greek and Arabic mathematical texts* 47 (May): 106–121. https://doi.org/10.1016/j.hm.2018.10.008.

Rashed, Roshdi. 2012. *Abū Kāmil, algèbre et analyse diophantienne. Édition, traduction et commentaire*. Berlin/Boston: Walter de Gruyter. https://www.decitre.fr/livres/algebre-et-analyse-diophantienne-9783110295610.html.

Rashed, Roshdi, and Bijan Vahabzadeh. 2000. *Omar Khayyam, the mathematician*. New York: Bibliotheca Persica Press.

Saidan, Ahmad Salim. 1978. *The arithmetic of Al-Uqlīdisī: The story of Hindu-Arabic arithmetic as told in Kitāb al-Fuṣūl Fī al-Ḥisāb al-Hindī*. Dordrecht: Springer. https://doi.org/10.1007/978-94-009-9772-1.

Sesiano, Jacques. 1993. "La version latine médiévale de l'algèbre d'Abū Kāmil." In *Vestigia Mathematica. Studies in medieval and early modern mathematics in honour of H. L. L. Busard*, edited by Menso Folkerts and Jan P. Hogendijk, 315–452. Amsterdam et Atlanta: Rodopi.

Woepcke, Franz. 1851. *L'Algèbre d'Omar Alkhayyámí. Publiée, traduite et accompagnée d'extraits de manuscrits inédits*. Paris: Duprat.

Chapter 4
Latin Middle Ages

Abstract This chapter discusses the way in which distributive-like properties appear in works belonging to the Latin medieval tradition of the *Elements*. This includes the *Liber Mahamaleth*, known for its direct influence on important works in the *algorismus* tradition, such as Fibonacci's *Practica Geometrie* and Sacrobosco's *Algorismus vulgaris*. The chapter also discusses the treatment of arithmetic rules in the works of Jordanus Nemorarius and Campanus de Novara.

Keywords Distributive-like rules · The Latin Euclidean Tradition · *Algorismus* tradition · Fibonacci · Sacrobosco · *Liber Mahamaleth* · Jordanus Nemorarius · Campanus de Novara

The Latin medieval authors that I discuss in this section drew their views on Book II and on the other distributivity-like results found in Euclid's *Elements* from a variety of sources. These include, in the first place, the various Latin versions of the *Elements*, but also various Arabic and Hebrew texts that were translated and circulated in Latin Europe, such as the *Liber Embadorum* (1145), Plato of Tivoli's Latin version of an Hebrew treatise by Abraham bar Ḥiyya (ca. 1065–1145). Bar Ḥiyya's treatise, *Ḥibbūr ha-meshīḥah we-ha-tishboret* (חיבור המשיחה והתשבורת), a title often translated as "Treatise on Measurement and Calculation", was a book on practical geometry (Bar Chijja 1912; Lévy 2001, 37–42; Sarfatti 1968, 64–128). Plato's Latin text introduced for the first time in Europe the techniques of Islamicate algebra for solving quadratic equations, thus antedating the first Latin translation of al-Khwarizmi's *Algebra*, by Gerard of Cremona (c. 1114–1187). The *Liber Embadorum* was widely read and it is known to have directly influenced the entire *algorismus* tradition (about which more is said below), and in particular Leonardo Fibonacci's treatises, including the *Practica* which I will discuss now (Curtze 1902, 5–7; Bar et al. 1931).

As already mentioned above, within Islamicate mathematics basic conceptions about numbers underwent significant changes relative to those originally underlying Euclid's work. These included the adoption and wide use of fractions and even irrational magnitudes in contexts where Euclidean propositions and methods were

© The Author(s), under exclusive license to Springer Nature Switzerland AG 2021   43
L. Corry, *Distributivity-like Results in the Medieval Traditions of Euclid's Elements*,
SpringerBriefs in History of Science and Technology,
https://doi.org/10.1007/978-3-030-79679-2_4

also invoked. These changes were later noticeable also in many Latin and Hebrew mathematical texts, such as those discussed below. The distributivity-like properties on which I focus here were handled very often as part of conscious attempts to clarify the foundations of arithmetic and to provide, for this mathematical field of knowledge, the kind of axiomatic foundations that geometry had enjoyed since the time of the *Elements*. In many cases, discussions related with distributivity-like properties played a central role in such attempts, and for this reason they are certainly worthy of close historical attention.

4.1   Liber Mahameleth

Liber Mahameleth is a text on commercial arithmetic, presumably written in or near Toledo around 1143–1153 (Corry 2013, 675–77). It is of particular interest for our discussion here because its preliminary section is explicitly devoted to presenting what the author saw as the foundational rules of arithmetic. These rules are presented from an original perspective that combines arithmetic and geometric considerations, and that betrays the kinds of concerns faced by an author trying to come to terms with the actual source of their validity.

The preliminary section of *Liber Mahameleth* comprises eighteen propositions: the first eight (LM-1 to LM-8)[1] are arithmetic in contents and style, while the last ten (LM-9 to LM-18) are adaptations of results adopted from geometry. The former comprise several results taken directly from Abū Kāmil's *Algebra*, including associativity of the product. The latter comprise results from Euclid's Book II. Proposition LM-9, which is the equivalent of Euclid's II.1, provides the basis for proving all the following ones.

It is evident that the author was aware that a reader with some knowledge of Euclid may have objected the order of the propositions and the reliance, for proving propositions that appear earlier in the original order of the *Elements*, on propositions that appear later on. Thus he wrote (Sesiano 2014, 597–98)[2]:

> We deemed it appropriate to add what Euclid stated in the second book, in order to explain with respect to numbers what he himself explained with respect to lines. It will be necessary for their proof to use certain propositions from the seventh book, for Euclid only spoke about

---

[1] In the critical edition of Vlasschaert (2010), the propositions are not numbered, but in the introductory text she provides algebraic renderings and numerates each corresponding formula. For ease of reference I am adding here a corresponding numeration for the propositions, with the initials LM, and I also refer to the corresponding page in the critical edition. A more recent, and in some senses more comprehensive, edition is that of Sesiano (2014). The latter had not been published when I completed (Corry 2013). Hence, in the interest of compatibility with my own previous article I will continue to refer here to the text as it appears in Vlasschaert (2010). Notice that Sesiano also uses a somewhat different numbering of the propositions.

[2] Sesiano comments in a footnote to this passage that, in spite of the warning, the only propositions which are used in the following proofs are VII.17 and V.24. On p. 593, footnote 56, Sesiano explains that actually VII.17 and VII.18 are sometimes interchangeably referred to.

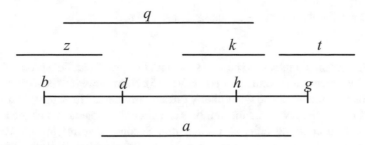

Fig. 4.1 *Liber Mahameleth* LM-9

numbers in the seventh book and the two following ones. For this reason Euclid should first be read and known thoroughly before embarking upon the present treatise on mahameleth.

Being this the case, the way in which the proposition is proved here is completely original and worth of attention. In the first place, the accompanying diagram (Fig. 4.1) is different from those appearing in any other medieval text for a version of II.1.

Like in Euclid's arithmetic books, the lines labeled with letters represent numbers but, unlike in Euclid, these numbers can be fractions or even roots. This means that arguments in which the units are counted and possibly rearranged will not work smoothly as was the case with Euclid or even al-Nayrīzī. The proof combines ideas relating to proportions of both numbers and continuous magnitudes, and makes crucial use of V.24. As we saw above, this proposition expresses a distributivity-like law for proportions. But at the same time, this proof of LM-9 also relies on VII.18, which connects the operations of multiplication and ratio formation for numbers. If we allow ourselves here a symbolic rendering, for the sake of brevity, the property embodied in VII.18 would be the following:

***VII.18**: b:c:ab:ac.*

The details of the proof of LM-9 are as follows:

In Fig. 4.1 above, two numbers a and bg are given, and bg is divided into parts bd, dh and hg. The proposition states that the product of a by bg equals the products of a by bd, a by dh and a by hg taken together. Further, q represents the product of a by bg, z the product of a by bd, k the product of a by dh, and t the product of a by hg. The proof can be rendered schematically with the symbolism of proportions (not used in the text, of course). It goes as follows:

(i.1) $a \cdot bg = q$; $a \cdot bd = z$; $a \cdot dh = k$; $a \cdot hg = t$;

(i.2)   By Euclid VII.18: $z{:}q :: bd{:}bg$; $k{:}q :: dh{:}bg$.

(i.3)   By Euclid V.24:[3] $z + k{:}q :: bh{:}bg$,

(i.4)   By Euclid VII.18: $t{:}q :: bd{:}bg$,

(i.5)   By Euclid V.24: $z + k + t{:}q :: bd + dh + hg{:}bg$,

---

[3] In the Vlasschaert edition, p. 26, there is no direct reference to V.24, but just "*sicut euclides dixit in quinto*". Sesiano (2014, 597–598) makes clear that the references are to VII.18 and V.24, as I indicate here.

(i.6)   But $bd + dh + hg = bg$, hence $z + k + t = q$.

QED.

Two important points should be stressed about this proof. Euclid's proof of VII.18 specifically depends on counting units (via VII.15). Here, however, the author applied it to segments, and he did so without any further comment. The segments represent numbers, but as already stated the numbers are not only integers and hence an argument based on counting units is problematic. Secondly, by relying on V.24, the author was actually grounding an arithmetic property on a property derived from the Eudoxian theory of proportions. Moreover, he did so via a result of Book II. On the face of it, once the author decided to rely on Book V, he might have followed the more straightforward approach of invoking V.1 (or one might also think of VII.5,6). Indeed, recall that V.1 embodies a law of distributivity of the product over the addition of several (not just two) magnitudes. But on closer look, multiplication in V.1 could be retrospectively seen now as concerning repeated addition of a magnitude to itself a number of times, whereas Euclid's II.1 (and hence also LM-9) is not meant as referring to that kind of multiplication. The "numbers" referred to in LM-9 are multiplied by another "number" rather than repeatedly added to themselves, as in V.1.

Also in this regard, then, the mixture of domains and approaches is quite unusual. It may well be the case that the author realized that the intended meaning of "multiple", when multiplying by segment a, required repeated reliance on V.24 rather than a direct application of V.1. But then again, this is somewhat roundabout since also VII.18 is applied and since the basic intention was to prove a property of numbers.

Upon examining the entire preliminary section more closely, one comes across an even more complicated picture, given that two special cases of distributivity-like properties are already proved in the book previous to LM-9. This is the case with the purely arithmetic propositions LM-6 and LM-7, which deal with the relationship between *division* and, respectively, subtraction and addition. Also the details of these proofs are quite interesting for reaching a clearer insight into the rationale of this preliminary section.

The first four propositions in the section comprise statements and proofs of elementary arithmetic properties such as associativity of multiplication or division of three or four numbers (with strong references to Abū Kāmil). The next proposition, LM-5 (pp. 20–22), is needed for proving LM-6. Its enunciation is the following:

> **LM-5:** If six numbers are given, such that the first is to the second as the third is to the fourth, and the fifth is to the second as the sixth is to the fourth, then that by which the first supersedes the fifth or is exceeded by the fifth is to the second as that by which the third supersedes the sixth or is exceeded by the sixth.

This is a modified version of V.24, when *subtraction* is involved instead of addition. As already mentioned, Robert Simson remarked that Euclid's proof of V.24 can be easily modified to obtain the result stated here in LM-5, namely by relying on V.17 rather than on V.18 (Simson 1804, 119). Indeed, this is precisely what the author of

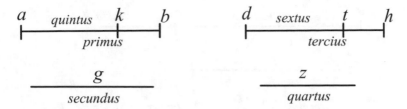

Fig. 4.2 *Liber Mahameleth* LM-5

Liber Mahameleth did in his proof. Let us consider the details, while referring to Fig. 4.2.

The six given numbers are *ab*, *g*, *dh*, *z*, *ak*, *dt*, and they define three proportions *ab*:*g* :: *dh*:*z*, and *ak*:*g* :: *dt*:*z*.

The proposition then states that *kb*:*g* :: *th*:*z*, and the proof can schematically be summarized in the following steps:

(j.1) from *ak*:*g* :: *dt*:*z* we obtain *g*:*ak* :: *z*:*dt*
(j.2) Now, from *ab*:*g* :: *dh*:*z and g*:*ak* :: *z*:*dt* it follows that *ab*:*ak* :: *dh*:*dt*
(j.3) From *ab*:*ak* :: *dh*:*dt* it follows that *bk*:*ak* :: *ht*:*dt*
(j.4) Finally since *ak*:*g* :: *dt*:*z* and *bk*:*ak* :: *ht*:*dt* it follows that *bk*:*g* :: *ht*:*z*

QED.

Some minor remarks on the proof are in order:

- The author justified step (j.1) by invoking Euclid's V.16, but from the latter proposition and from *ak*.*g* :: *dt*:*z* what we obtain is *ak*:*dt* :: *g*:*z*.
- Steps (j.2) and (j.4) rely on V.22, which is not mentioned by the author.
- The crucial step is (j.3), which, as Simson said, depends on V.17. The author wrote here "*Secundum proportionalitatem*".
- Notice that both V.17 and V.22 are proved in Euclid directly from the Eudoxean definition of proportion. There are no parallels to these two propositions in Book VII. Thus, also in this sense the author is relying here directly on Book V for his foundational result.
- Whereas in the enunciation, the possibility "or is exceeded" (by the fifth/sixth) is mentioned, there is no reference to that in the proof.

The proof of LM-6 (pp. 22–23), which handles *subtraction* rather than addition, is a direct application of LM-5.

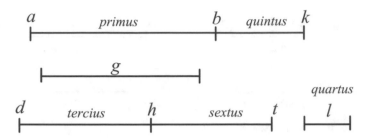

Fig. 4.3 *Liber Mahameleth* LM-7

Referring once again to the above diagram, two numbers are given, *ab* and *ak*, *kb* being their difference. When divided by *g*, they yield, respectively, *dh* and *dt*, and *th* is their difference. The proposition thus states that if the difference *kb* is divided by *g*, the result is *th*, the difference of the divisions.[4] The argument of the proof is simple: from *ab/g = dh* it follows that *dh · g = a* and hence *g:ab* :: 1:*dh*. Likewise, one can deduce *g:ak* :: 1:*dt*. Here one can apply LM-5, from whence *g:kb* :: 1:*th*, and from here, the desired results follows easily.

The author indicated that the following proposition is similar, but now for the case of addition (rather than subtraction as in LM-6). Therefore one needs a rule that is similar to LM-5, but applies to addition. But such a rule, he stated, had been already proved by Euclid and hence it would not be necessary to prove it in the treatise again. The reference is, of course, to V.24. Thus, LM-7 is formulated as follows (p. 24):

> **LM-7:** When any two numbers are divided by another number, then the outcomes of both divisions taken together equal the result of dividing by the same divisor both numbers taken together.

Referring to the diagram reproduced in Fig. 4.3, the proposition states that if we divide *ab* by *g* and the result is *dh*, and if we divide *bk* by *g* and the result is *ht*, then, the result of diving *ak* by *g* is *dt*. The proof is similar to that of LM-6, but in its crucial step it relies directly on Euclid's V.24.

Liber Mahameleth comprises many results that are illustrative of the challenges faced by medieval authors trying to come to terms with the very idea of providing foundations for arithmetic. Distributivity-like results are among such results. Of particular interest is the way in which an underlying tension manifests itself, arising

---

[4] On p. 23 of the Vlasschaert critical edition there is an additional diagram. It does not seem, however, to fit the argument presented in the proof.

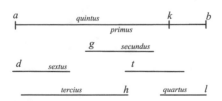

from the attempt to understand the basic properties of numbers without thereby giving away the traditional, Euclidean centrality of geometry as the field that is better understood and axiomatically founded.

4.2   Fibonacci: Geometry and Arithmetic

Another illuminating perspective on the use of distributivity-like properties in Latin texts that I want to discuss here concerns two relatively minor works of Leonardo Fibonacci (1170–1240/50): *De Practica Geometrie* completed around 1220 and *Liber Quadratorum* completed in 1225.[5] Alongside his broad knowledge of Islamicate mathematics, Fibonacci was well acquainted with the Euclidean text and he mastered the techniques taught there. Still, it is not completely certain what version of the *Elements* was available to him. He may have simultaneously relied on several of the existing translations, including the ones made directly from Greek (Folkerts 2004, 109–10; Hughes 2008, xix). Abū Kāmil's treatise on algebra, which Fibonacci may have known either directly or indirectly, was also a main influence for him. But it seems that the main influence on Fibonacci when working on the *Practica Geometrie* came from bar Ḥiyya's *Liber Embadorum* (Hughes 2008, xxiii–xxiv).

The *Practica* teaches its readers how to measure areas of any type, and its main target audience was that of craftsmen of various skills.[6] One important difference between this treatise and Fibonacci's more famous *Liber Abbaci* is found in an entire section where Fibonacci cites twelve propositions from Euclid's *Elements*, the first nine of them being versions of propositions stemming from the first part of Book II (II.8 is absent). He also provided proofs for some of these propositions which differ interestingly from Euclid's or Heron's (Corry 2013, 677–684). What is of special interest for our account here is that in all of these propositions, the kind of distributivity afforded by II.1 (or its direct derivatives, such as II.2,3) plays a central role. Fibonacci presented these results in two different versions. First, an arithmetic version that applies to "any number" and is proved by counting units. Secondly, a more geometric version in which "a straight line" is divided into segments as in II.1. For the second kind of statements, he typically provided no proof. It is worth examining in detail some instances of this.

Proposition 31 of the *Practica* (PG-31) is an arithmetic version of II.1. The numbers are represented over a line, and they are meant to be arbitrary, but at the same time, specific numerical values appear in the diagram (Fig. 4.4). Along the proof, the character of the numbers and the operations with them move freely and constantly, back and forth and without any further comment, between the arithmetic and the geometric context. Thus, the number ab is given the value 10, and it is divided

---

[5] I thank Stela Segev for indicating me the relevance of discussing the *Liber Quadratorum* in this section.

[6] See (Moyon 2012), for a broader discussion of the genre of *Practica Geometriae* and its relation with the early development of algebra.

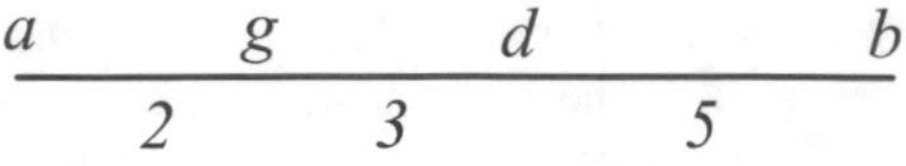

Fig. 4.4 Fibonacci's
diagram for the arithmetic
version of Euclid's II.1

$$
\begin{array}{cccc}
a & g & d & b \\
\overline{} & \overline{} & \overline{} & \overline{} \\
2 & 3 & 5 &
\end{array}
$$

into *ag*, *gd*, and *db*, with values 2, 3, and 5, respectively. Fibonacci claims that the products of *ag* by *ab* together with *gd* by *ab* and *db* by *ab* equal the product of *ab* by *ab* (Hughes 2008, 26–27).

The reason adduced is, simply, that "the number of units in part *ag* with those in *ad* will produce the product of *ag* by [*ad*]."[7] And the same is said for the other two pairs. Hence, Fibonacci concludes, "because there are as many units in the number *ab*, namely in the parts *ag*, *gd*, *db*, so many are united in the number *ab* from the multiplication of *ag*, *gd* and *db*, by *ab*." And on the other hand "as many units as there are in *ab*, so many arise from the multiplication of *ab* by itself". His argument, then, is similar to that found in Euclid's VII.5. After concluding that both products are equal, as required, Fibonacci also gave a numerical example.

In PG-32, Fibonacci divides a "straight line" into many parts and multiplies each part by "another line", and then states that "the sum of all the products equals the product of the whole divided line by the other line". He does not provide a proof for this claim, but just a numerical example: 2, 3, and 5 multiplied by 12. PG-33 is a version of II.3 and PG-34 of II.4. In both cases the enunciation is for "straight lines", rather than for numbers, and the proof is based on the previously proved results on distributivity. In both cases Fibonacci added that the result "can be shown with numbers".

Fibonacci's skills as a theoretician are manifest, perhaps more than in any of his other texts, in the *Liber Quadratorum*. The main motivation behind the entire treatise is Fibonacci's attempt to solve a challenge posed to him by John of Palermo, a scholar working in the court of Emperor Frederick II (1194–1250): to find a square number from which, when five is added or subtracted, the result is always a square number. The treatise, which was known for example to Tartaglia, but fell into oblivion until the end of the eighteenth century, is clearly influenced by the Diophantine tradition of indeterminate analysis, which Fibonacci knew as it had developed in the hands of Islamicate mathematicians. It comprises both propositions and problems, and the text is written with no clear and explicit break from one result to the next. There is a mixture of numerical examples and discussions about specific cases, sometime before and sometimes after the relevant proposition or proof, with no clear structure to that. At the same time, the proofs continually mix arithmetic with geometric arguments and representations of the numbers involved.[8]

---

[7] In Hughes' translation, in this specific part of the proof, *ab* and *ad* are interchanged in two places, which I corrected here, to make them appear as in the original Boncompagni edition (1856, 27). Likewise, Fibonacci's conclusion, that follows immediately, sounds as a somewhat weird type of mathematical reasoning, but it is indeed what the original says.

[8] An edited version of the treatise, based in a manuscript found in the Ambrosian Library in Milan appears in Boncompagni (1856). There is an English translation by L.E. Sigler (Fibonacci 1987), where the propositions are numbered, and the same numbering is also followed in Katz et al. (2016, 112–16). In this counting, the treatise comprises twenty-four propositions in all, of which eight

Fig. 4.5 Fibonacci's
diagram for LQ-2

$$a \qquad\qquad b \qquad\qquad d \quad g$$

The problem lying at the heart of the treatise is solved in proposition LQ-17. Interesting as it is, it does not directly touch upon the issue that concerns us here. It is important to notice, however, that in order to solve that problem, Leonardo introduced the idea of what he called "congruous numbers" (*numerus congruus*). Thus, if two relatively prime numbers n, m are both odd, then Fibonacci proved (LQ-12) that the product $nm(n + m)(n - m)$ is a multiple of 24, whereas in case that one of the numbers is even then $4{\cdot}nm(n + m)(n - m)$ is a multiple of 24. In both cases he called such products "congruous numbers". In LQ-14, Fibonacci showed how to find a number that, when added to a square and subtracted from the same square, yields squares. Symbolically, the problem is to find three square numbers x^2, y^2, x^2, and a number c such that:

$$x^2 + c = y^2 \text{ and } y^2 + c = z^2.$$

If found, it turns out that, as required, both adding c to y^2 and subtracting c from y^2 yield squares. It also turns out, as part of the proof, that c is a congruous number. The proof of this proposition, which is crucial for the proof of LQ-17, is long and extremely involved, to the extent that it would be beyond the scope of the present account to present it here.[9] At the heart of the proof lie two "identities", which are certainly distributive-like in spirit, and which can be symbolically rendered as follows:

$$(n^2 + m^2)^2 - 4nm(n^2 + m^2)^2 = (m^2 + 2nm - n^2)^2,$$
$$(n^2 + m^2)^2 - 4nm(n^2 + m^2)^2 = (n^2 + 2nm - m^2)^2.$$

For our purposes here, I find it more convenient to discuss in detail two of the proofs, where the reliance on distributive-like properties appears in a more visible and significant manner. I start with LQ-2 (Fig. 4.5), which reads as follows:

LQ-2: Any square number exceeds the square immediately before it by the sum of the roots.

Here ab and bg are two successive numbers, with one of them being bigger than the other by 1. Thus bg is one plus ab, and if we subtract from bg the unit dg, the remainder bd equals ba. In the proof, Fibonacci relies on propositions taken from Book II of the *Elements*, which he phrases for numbers, and without specifically

---

are theorems and fourteen are problems. A slightly different numbering and counting appears in McClenon (1919). No such numbering, however, appears at all in the Boncompagni edition. For reasons of convenience I follow here the numbering of Sigler, but for the sake of accuracy I prefer to refer directly to the Boncompagni edition.

[9] See (Fibonacci 1987, 53–74). I intend to devote a separate article to discuss this, as well as other proofs in the treatise.

Fig. 4.6 Fibonacci's
alternative diagram for LQ-2

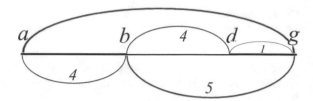

mentioning which proposition he refers to. I indicate them here in describing the argument schematically, which goes as follows:

(k.1)   By II.4: $(bg)^2 = (bd)^2 + (dg)^2 + 2 \cdot (bd \cdot dg)$,
(k.2)   But $bd \cdot bd = ab \cdot ab$, and hence $(bg)^2 = (ab)^2 + (dg)^2 + 2 \cdot (bd \cdot dg)$,
(k.3)   But since $dg = 1$, $dg \cdot dg = 1 = dg$, and $2 \cdot (bd \cdot dg) = 2 \cdot bd = ad$,
(k.4)   Hence, $(bg)^2 - (ab)^2 = dg + ad$.
(k.5)   But $dg + ad = ag = bg + ab$.

Hence, "the square built on the number bg supersedes the one built on ab by the quantity of the addition of their roots, which are ab and bg." (p. 60).

QED.

But this is followed by a second, alternative argument, which refers to the diagram in Fig. 4.6.[10]

Here the argument is much simpler and shorter than the previous one, and indeed it follows from a straightforward, tacit reliance on Euclid's II.6. Indeed, the diagram embodies the conditions of II.6, namely, "the number bd equals number ba, since the entire ad is divided into two equal parts at point b, to which it is added the unit dg." It follows that "the multiplication of dg by ag, together with the square built on the root ab equals the square built on the root bg." Symbolically this can be rendered as:

$$dg \cdot ag + (ab)^2 = (bg)^2.$$

And as in the first argument, since dg is 1, this expression implies that "the square of bg exceeds the square built on ab, by the addition of their roots, which added together yield the number ag." Or, symbolically:

$$(bg)^2 - (ab)^2 = 1 \cdot ag = bg + ab.$$

And to this, Fibonacci adds yet a more general conclusion, namely, that any square exceeds any smaller square by the product of the difference of the roots by the sum of the roots or. Symbolically stated this is a truly distributive law of multiplication:

---

[10] The full proposition is presented in Boncompagni (1856, 60–62), and this diagram is referred to as Fig. 10. In Sigler's translation, it is not at all reproduced, and the reference is made to the same line $abdg$ as in the first part of the proof. In fact, the numerical values added in this diagram are not needed for presenting the alternative argument, but it seems indeed important to stress the use of two different diagrams and the fact that the second one also comprises numerical values.

$$a \qquad\qquad g \qquad\qquad d \quad\;\; b$$

Fig. 4.7 Fibonacci's diagram for the generalization of LQ-2

$$n^2 - m^2 = (n + m)(n - m).$$

Moreover, the previous result is a particular case of this one, since there the difference between the two roots of the consecutive squares is 1. Fibonacci proves this more general result, while referring to the diagram in Fig. 4.7.

The proof can be rendered symbolically as follows:

(1.1) $gb = ag + db$, [i.e. $gd = ag$],
(1.2) Hence by II.6, $(gb)^2 = (ag)^2 + (db \cdot ab)$,
(1.3) But $db = gb - ag$ and $ab = gb + ag$,
(1.4) Hence $(gb)^2 - (ag)^2 = (gb - ag) \cdot (gb + ag)$.

QED.

Another relevant result of the treatise, worthy of consideration here is the following:

LQ-19: To find a square number for which both the sum of itself and its root and the difference between itself and its root are square numbers.

In other words, it is requested to find three numbers, n, m, p such that:

$$n^2 - n = m^2 \quad \text{and} \quad n^2 + n = p^2,$$

As before, Fibonacci constructs the numbers by reference to a geometric diagram (Fig. 4.8), representing three square numbers ab, ag, ad and their congruous number bg, which is also gd.

We learnt from LQ-14 that it is possible to build such numbers, and they satisfy the conditions:

$$ag - bg = ab \quad \text{and} \quad ag + gd = ad.$$

The numbers ez, ei, eh in the diagram, represent the result of dividing the square numbers by the congruous number bg (or gd). Hence we obtain:

$$(ag/bg) - 1 = (ab/bg) \text{ and } (ag/gd) + 1 = (ad/gd),$$

$$\text{or } ei - 1 = ez \text{ and } ei + 1 = eh.$$

Now, on ei build a square ek (hence $el = ei$), complete the rectangle lh, and then at point z, raise a line zt parallel to ik and el. Also, by construction, $zi = bg/bg = 1$ and $ih = gd/gd = 1$. Hence $zi = ih = 1$.

Based on these construction, a schematic rendering of Fibonacci's reasoning is the as following:

(m.1)   The area of the rectangle *kh* equals $1 \cdot el$, and $el = ei$,

(m.2)   Hence the area of *kh* (or of *kz*) equals the root of the square ek,[11]

(m.3)   Hence, $ek + kh = lh$ (*kh* being the root of the square ek, namely el),

(m.4)   Hence $el^2 + el = el \cdot eh = lh$.

(m.5)   Likewise $ek - kz = zl$ (*kz* being also the root of the square ek, namely el).

(m.6)   Hence, $el^2 - el = el \cdot ez = et$.

(m.7)   But since $ag/bg = ei$ and $ab/bg = ez$,[12] hence $ag{:}ab :: ez{:}ei$, and since ab, ag, are square numbers, hence ez, ei hence the ratio $ez{:}ei$ is that of two square numbers. Hence the product $ez \cdot ei$ is a square number, and so is the area of et (since $ei = el$).[13]

Fig. 4.8 Fibonacci's diagram for problem LQ-19

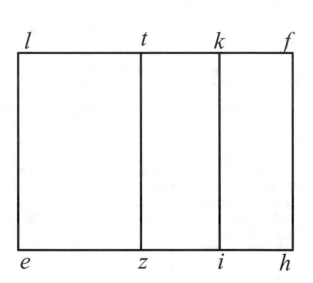

[11] Boncompagni (1856, 99–100): "… superficies itaque *kh*, vel *kz*, est radix tetragoni *ek*. Ergo, super tetragonum *ek* addatur eius radix, scilicet superficies *kh*, proveniet superficies *lh*; et si ex quadrato *ek* ausseratur eius radix, quiest *kz*, remanebit superficies *zt*." This means an equality between two numbers, which are represented in the diagram one as a line and the other as an area.

[12] Of course, this is not formulated as an operation involving symbolic terms that allow for formal manipulation of the numbers. Rather it reads as follows (Boncompagni 1856, 99): "… et dividatur unuquisque quadratorum *ab*, *ag*, *ad* per congruum *bg*, et proveniant numeri *ez*, *ei*, *eh* …".

[13] Boncompagni (1856, 100): "Est ergo sicut *ab* ad *ag*, ita *ez* ad *ei*; sunt enim quadrati numeri *ab* et *ag*: ergo proportion numeri *ez* ed *ei* sicut proportion quadrati numeri ad quadratum numerum. Quare ex ductu *ez* in *ei* proveniet quadratus numerus. Sed *ei* recte equalis est recta *zt*, cum sit equalis recte *ik*, tetragonum enim est superficies *ek*: ergo superficies *et* est quadratus numerus."

(m.8)   Similarly, since $ag/gd = ei$ and $ad/gd = eh$, hence $ag{:}ad{:}ei{:}eh$, and since ag, ad, are square numbers, hence the ratio $ei{:}eh$ is that of two square numbers. Hence the product $ez{\cdot}ei$ is a square number, and so is the area of lh (since $ei = el$).

And thus—Fibonacci concluded—a square number ek has been found, as requested, to which by adding its root, which is kh, a square number is obtained, which is the number lh. And if from the square ek its root ei is subtracted, the remainder is a square number et. QED.

Two points are worthy of emphasis concerning this proof. First is the argument behind steps (m.7)–(m.8), which seem to rely on a combined use of two propositions taken from Euclid's *Elements*, namely VII.17 and VIII.24. They read as follows:

> **VII.17:** if a number multiplied by two numbers makes certain numbers, then the numbers so produced have the same ratio as the numbers multiplied.
>
> **VIII.24:** if two numbers have to one another the ratio which a square number has to a square number, and the first is square, then the second is also a square.

Fibonacci, to be sure, neither mentioned the propositions nor gave the details that follow, but it seems evident that this is how he may have justified the steps:

(m.9)   On the one hand, from VII.7, we obtain $ei{:}eh :: ei{\cdot}eh{:}eh{\cdot}eh$,

(m.10)   on the other hand, we have $ag{:}ad :: ei{:}eh$, and hence $ag{:}ad :: ei{\cdot}eh{:}eh{\cdot}eh$;

(m.11)   but since ag, ad, and $eh{\cdot}eh$ are all square numbers, it follows from VIII. 24, that also $ei{\cdot}eh$ is square, as claimed by Fibonacci.

The use of Euclid's theory of proportions in this manner is both ingenious and necessary, since, as indicated above, the text does not involve any kind of formulation that allows for operating on the numbers, other than with the help of propositions of this kind.

Strongly related with this is the second point that is particularly important for our account here, namely, the way that distributivity-like properties appear on the argument. So, on the one hand Fibonacci applies, in a rather straightforward manner, a distributivity law for division when passing [via $(ag/bg) - 1 = (ab/bg)$] from $ag - bg = ab$ to $ei - 1 = ez$. In the text this is formulated as follows (p. 99):

> … and since the number ez is obtained by dividing the number ab by bg, and the number ei is obtained by dividing the number ag by the congruous, the number zi is obtained by dividing bg by itself and hence zi equals 1.

On the other hand, the aim of the argument is to attain a distributivity-like result involving the multiplication of the square numbers in order to obtain (m.4) and (m.6). Part of the argument could be skipped if Fibonacci had thought of applying directly a distributivity argument as he did for division. Indeed, given that $ei - 1 = ez$, and that $ei = el$, if we multiplied all terms in the identity by el, we would directly obtain $el^2 - el = et$, which is (m.6), and the same could be done for (m.4). Instead, Fibonacci had to go around and apply some additional steps that combine geometric and arithmetic reasoning.

As we see, then, in these two texts, like in some others of him, Fibonacci's state-ments of the propositions allowed considering them simultaneously as related to both geometry and arithmetic, and he could move quite freely from one realm to the next when necessary. Some of the geometric results he relied upon he took for granted, some he just illustrated with numerical examples. He introduced new proofs for some of the Euclidean propositions but also used such canonical propositions in new contexts. And within this account, distributive-like results and rules appear in different ways in these two treatises: sometimes he relies on them explicitly (refer-ring to the first propositions of Book II), sometimes he uses them implicitly and perhaps inadvertently, and sometimes he developed roundabout arguments aimed at establishing results of that kind.

4.3 Sacrobosco: *Algorismus*

Johannes de Sacrobosco (ca. 1195–1256), who taught at the University of Paris, is best known for his book on astronomy, *Tractatus de Sphaera*. Written around 1230, it became the most successful and widely read treatise used to teach the basics of the geocentric picture of the world, based mainly on Ptolemy's *Almagest*, but also on ideas stemming from Islamicate astronomy. First in manuscript version and then, beginning in 1472, in a large number of printed editions, it became the book of choice and required reading in European universities well into the seventeenth century (Valleriani 2020). Several years previous to the *Sphaera*, Sacrobosco completed a no less interesting and influential treatise, *Tractatus de Arte Numerandi*, or, as it is sometimes also called, *Algorismus vulgaris*. Some ideas appearing in the *Algorismus* are relevant to our discussion here.

The medieval *algorismus* tradition was devoted to teaching the use of Hindu-Arabic numerals, how to write them and how to calculate with them, including calculations involving sexagesimal and common fractions. The name *algorismus* derived from a Latinization of the name of al-Khwārizmī, typically appearing at the beginning of the tracts, even though the historical connection with the author, from whose original texts much of this knowledge originated, was not understood. "A certain philosopher named Algus", explained Sacrobosco, "wrote this brief science of numbering, for which reason it is called *Algorismus*."[14]

A large number of Latin *algorismus* treatises were written and circulated between the twelfth and fifteenth century. They provide the main evidence documenting the genesis, initial adoption and expansion of the decimal arithmetic based on the use of ten digits, including zero (Allard 1991). Together with *Carmen de Algorismo*, by the French Alexander de Villa Dei (d. ca. 1240), Sacrobosco's tract was one of two such texts, which, written in the first half of the thirteenth century, became widely used and influential (Ambrosetti 2016). It was copied in hundreds of manuscripts and later printed, and went through twelve editions within 33 years. Both treatises are limited

---

[14] Grant (1974, 94). Grant translated the text from (Curtze 1897).

to calculations with integers, and they present their results by briefly describing each step of the various calculations. One important feature that distinguishes the *algorismus* tradition from that of the abacists, is that it typically comprises explanations about what is seen as *seven* basic operations: addition, subtraction, duplication, halving, multiplication, division and root extraction. Typically, no proofs are given for the rules governing these operations. Numerical examples appeared in words, rather than with digits (Folkerts 2001, 17–19). It is likely that, Sacrobosco's *Algorismus* was the Latin text of preference used for the learning of Hindu–Arabic numerals and related procedures as part of the European university quadrivium curriculum.

Sacrobosco's explanation of the general case of multiplying two numbers whatsoever, written with the positional system based on ten symbols, is indeed an explicit formulation of the distributivity law of multiplication over addition, though in a somewhat restricted sense that he does not go on to generalize. Sacrobosco starts by defining three kinds of numbers as follows: *digits* (i.e., any number smaller than 10), *articles* (multiples of ten), and *composite numbers* (i.e., a mixture of the previous two kinds). There are six rules for multiplication. The first two explain how to multiply, respectively, a digit by a digit and a digit by an article. Rule four explains how to multiply a digit by a composite number, namely, by applying a distributivity principle (Curtze 1897, 98):

> The multiplier (*multiplicans*) must be multiplied into every part of the composite number—the digit by the digit in accordance with the first rule; the digit by the article in accordance with the second rule. The products (*producta*) are added and the sum (*summa*) of the whole multiplication will be obvious.

The fifth and the sixth rules expand this to the multiplication of an article by a composite number and then, finally, of any composite number by any composite number, which is done as follows:

> When a composite number multiplies a composite number, each part of the multiplier must be multiplied by each part of the multiplying number, and the products are added: the sum (*summa*) [of the products] will then be obvious.

This way to operate in the decimal system with Hindu-Arabic numerals permeated an implicit use of distributive-like arguments at the core of arithmetic. Given the ubiquity and influence of *algorismus* treatises in the Latin and, as we will see below, also the Hebrew medieval traditions, it is then obvious that such arguments constituted a basic component of mathematical practice. As in some of the examples already seen and more to be seen below, the constant presence and use of distributive-like ideas sometimes led to explicitly formulating them, and sometimes to just use them in ingenious, yet tacit and perhaps even inadvertent ways.

4.4 Jordanus Nemorarius: *Arithmetica*

The two most prominent medieval authors who were involved with questions pertaining to the foundations of arithmetic were Jordanus Nemorarius and Campanus de Novara. Both were active in the thirteenth century. Because of the originality of their ideas and the influence they exerted on mathematicians of later periods of time, they deserve a more detailed, separate discussion concerning the ways in which distributivity-like properties appear in their works. The interested reader may find such a detailed discussion in Corry (2016). For the sake of completeness in the presentation here, I summarize now the most important points related to it.

About Jordanus we only know that he lived before 1260. His treatise *Arithmetica*, was perhaps the most important medieval treatise on the topic. In it, Jordanus sought to achieve for arithmetic what Euclid had done for geometry, in all what concerns the derivation of the body of arithmetic from definitions, postulates, and common notions. Moreover, he explicitly avoided reliance on geometrical concepts or results of any kind when putting forward his project (Corry 2013, 684–689). A central role was accorded in this treatise to five distributivity-like results, respectively parallel to Euclid's VII.5, VII.6, V.1, and (in two different versions) II.1. It is quite straightforward to establish the following parallels:[15]

Jordanus' *Arithmetica*	Euclid's *Elements*
A-I.4	VII.5
A-I.5	VII.6
A-I.6	V.1
A-I.9	II.1
A-I.10	II.1*

The first three of these are, like their Euclidean counterparts, statements of the type "if … then". For example:

A-I.4: If a first number is the same part of a second number that a third is of a fourth, then the first and the third are the same part of the second and the fourth that the first is of the second.

The last two of these propositions (A-I.9,10) embody a property similar to left- and right-distributivity of the product over addition for numbers. In order to convey the overall feeling of Jordanus' style in handling these propositions, it seems convenient to discuss them in some detail. They are the following:

A-I.9: That which is obtained by multiplying any number by as many as one pleases equals that which is obtained by multiplying the same [number] by their combination [i.e., their sum].

A-I.10: That which is obtained by multiplying as many numbers as one pleases by some number equals that which is obtained by multiplying their combination [i.e., their sum] by the number.

If we allow ourselves a symbolic rendering of the two for the purposes of illustration we obtain the following:

---

[15] For easiness of reference I use here a numeration of the propositions which does not appear in the original.

A-I.9: $a \cdot b + a \cdot c + a \cdot d + \cdots = a \cdot (b + c + d + \cdots)$,

A-I.10: $b \cdot a + c \cdot a + d \cdot a + \cdots = (b + c + d + \cdots) \cdot a$.

Notice that whereas A-I.10 embodies a legitimate arithmetic operation (a sum of numbers being multiplied by another number) a parallel proposition stated in the purely geometrical context of Euclid's Book II would make little sense. Indeed, such a parallel formulation would amount to something similar to the following:

II.1': If there are two straight lines, and one of them is cut into any number of segments whatever, then the rectangle contained by the two straight lines equals the sum of the rectangles contained by each of the segments and the uncut line.

It would not be clear, in the first place, what is the sum of rectangles that such a formulation would involve, given that in the context of Book II the sums of rectangles are concatenations, one to the right of the previous one. Moreover, to the extent that one can make sense of this formulation, it does not seem to add anything of real content to the original II.1.

Jordanus decision to include A-I.10 is even more remarkable given that in A-I.8 he had proved the commutativity of the product in general:

A-I.8: If two numbers are multiplied alternately, the same number is obtained in both cases.

Why did Jordanus nevertheless prove A-I.10, then, if it follows from applying A-I.8 to A-I.9? The details of the proof clearly indicate that propositions A-I.9,10 actually represent not just two different versions of the same situation, but actually two propositions that were truly different for Jordanus: a "number" is multiplied by "as many numbers as one pleases", and "as many numbers as one pleases" are multiplied by "a number" (Corry 2016, 322–323).

It is interesting to notice that whereas the counterparts of the first two of Jordanus' propositions, A-I.4,5 (i.e., V.5,6) had originally been formulated by Euclid for numbers, the one which is parallel to A-I.6 (i.e., V.1) was a proposition about ratios. Euclid's II.1, in turn, was about areas of rectangles. In the *Elements*, each of them was formulated in accordance to the topic of the Euclidean book where they were treated. Accordingly, Euclid's proofs relied on different kinds of arguments corresponding to the various different contexts. But in stark contrast to this, Jordanus presented all these five propositions in the introductory section of his treatise as purely arithmetic properties satisfied by operations with numbers. Remarkably, the proofs of these propositions, while fully arithmetic in essence, retain to some extent the taste of the different context (e.g., the geometric one), where Euclid had originally formulated them.

In addition to these five propositions, the preliminary section of the *Arithmetica* also contain additional cases of distributivity-like properties of various kinds. These are, as a matter of fact, versions of II.1–3, that either generalize or provide particular cases of the previous ones, and that are proved while relying on those previous ones, particularly on A-I.9. Thus, for instance, the following three (of which I just give the symbolic rendering):

A-I.11: $(a + b + c + \cdots) \cdot (p + q + r + \cdots) = a \cdot p + a \cdot q + a \cdot r + \cdots + b \cdot p + b \cdot q + a \cdot r + \cdots + c \cdot p + c \cdot q + c \cdot r + \cdots$

A-I.13: If $a = b + c + d + \cdots$ then $a \cdot a = a \cdot b + a \cdot c + a \cdot d + \cdots$

A-I.14: If $a = b + c$ then $a \cdot b = b \cdot b + b \cdot c$.

We can summarize this account by stating that Jordanus' conscious attempt to provide a rigorous presentation of arithmetic, not found in previous treatises, led him to make a clear distinction between repeated addition of numbers and multiplication of two numbers, but at the same time to present these two ideas as closely related. A main focus of attention in pursuing that distinction appears in relation with distributivity-like results, and a possible reason for this is that in those treatises where he learnt his arithmetic, he did not find a satisfactory treatment of such results.

4.5   Campanus: *Elements*

Campanus de Novara (1220–1296) published a Latin version of the *Elements* around 1260. His treatment of Book II does not differ from the standard Euclidean one found in other medieval versions of the treatise (Corry 2013, 689–692). His treatment of distributivity-like results in the framework of Books V and VII, however, is highly original. In the preliminary section to Book V, Campanus introduced lengthy additions and comments. He did not hesitate to explain to his readers what, in his view, was in Euclid's mind when writing this or that definition. Campanus also commented on the highly difficult character of the theory of proportions as presented in Book V, while stressing explicitly that these difficulties arise mainly from the need to deal, within one and the same framework, with irrational as well as with rational ratios (Busard 2005, 173–75). As I already explained in (Corry 2013), Campanus devoted some efforts to discuss, echoing Jordanus, the significance of Euclid's double treatment of proportions, once for continuous magnitudes and once for numbers.[16]

In order to help the reader following and understanding the arguments of the proofs in Book V, Campanus associated numbers to the segments that appear in the diagrams (he did the same in many diagrams of Books VII–IX as well). Jordanus had followed a similar approach, and it was quite natural for him to do so as part of his treatment of results in the arithmetical books. It seems much less natural, however, to find this arising in the case of Campanus, given his stress on the essential difference between handling proportions that involve continuous magnitudes and those that are purely arithmetic. Let us see how this appears, for instance, in the diagram accompanying V.1 (Fig. 4.9).

Here the three magnitudes a, b, c are said to be equimultiples of d, e, f respectively, and the proposition states that $a + b + c$ is the same equimultiple of $d + e + f$ as a is of d. The numbers appearing in the diagram are not even mentioned in the

---

[16] Campanus' discussion of proportions also raises some additional issues concerning both textual and conceptual difficulties, but they are beyond the scope of this article. See (Rommevaux 2007).

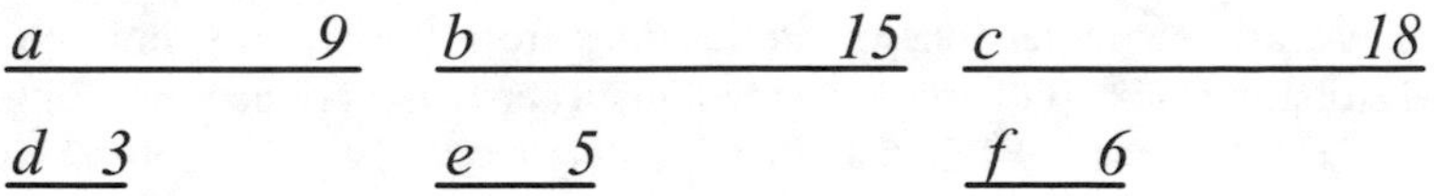

a	9	b	15	c	18
d	3	e	5	f	6

Fig. 4.9 Campanus' diagram for Euclid's *Elements*—V.1

specification or in the proof, but one can imagine that they may have helped the reader follow the argument.

Also in the introductory section to Book VII Campanus devoted focused attention to distributivity-like properties. Thus, one of his common notions, which is fundamental to the attempted systematic foundation of arithmetic, states that if the unit is multiplied by any number or if a number is multiplied by the unit, then the result is the number itself (Busard 2005, 231). Campanus added three additional ones, which I want to include here under the category of "distributivity-like". They are the following:

- Any number that measures two numbers, measures also their sum.
- Any number that measures some number, measures also any number measured by it.
- Any number that measures the whole and the deducted, measures also the remainder.

Intrinsically related with the attempt to provide an axiomatic foundation for arithmetic is the idea of the autonomy of such a body of knowledge vis-à-vis the other parts of mathematics presented in the *Elements*. Campanus consistently stressed this issue throughout the arithmetic books, and in particular he stressed the autonomy of proofs in Book VII vis-à-vis those of Book V. As I already indicated in Corry (2013, 325–326), Campanus explained that the "*propria principia*" of the two books are different and, hence, corresponding propositions should be proved separately, and based on those specific principles alone in each case. In relation with the proof of VII.5, for instance, we find the following statement (p. 235):

> Euclid wanted that the arithmetical books would not have to rely on the previous ones, but rather that they would stand by themselves, and results that he proved in the fifth books for quantities in general he proved here for numbers in this fifth of the seventh.

When examining the proofs in some detail, however, we notice that in some cases this autonomy did not go beyond repeating, while fully rewording for numbers, an argument already presented in Book V.[17]

The most important point to mention here in relation with Campanus' version of the *Elements* concerns a collection of fifteen commentaries added after the proof of IX.16. As already explained in Corry (2013), these commentaries comprise, among

---

[17] In other Latin versions of the *Elements*, instead of such a repetition, often there is just a direct reference to a corresponding proposition in Book V. See Busard (2005, 560).

other things, arithmetic versions of propositions from Book II that embody interesting distributivity-like properties. As mentioned above in §2.4, an interesting peculiarity of Euclid's proof of IX.15 is that it transgressed the self-imposed rules of separation between realms, and used arithmetic versions of II.3,4, without further comment. Al-Nayrīzī in his own commentary, in turn, formulated IX.16 as a generalized version of IX.15, and added in relation to it his own arithmetic versions of II.1–4. Now Campanus, following on al-Nayrīzī's footsteps, also added here his own commentaries.

A reader of Campanus who was also acquainted with Jordanus' *Arithmetica* (if there was any) would have easily recognized the close relationship (sometimes verbatim repetition) of Campanus' comments and Jordanus' basic rules of arithmetic discussed above. Book VII opened by rehearsing Jordanus' attempt to providing an axiomatic foundation, yet Campanus was also using the opportunity to include those elementary propositions that Jordanus had developed in the first chapter of his book. But as with Book VII, some of the technical changes that Campanus introduced in his presentation lead to some noteworthy differences. Thus, for example, Campanus' first two comments state two symmetric, distributivity-like rules involving products and additions. Their enunciations are parallel to, but not identical with, Jordanus' A-I.9 and A-I.10. This is how they appear in the text:

C-IX.9-1: That which is made by multiplying a number by as many as we wish equals that which is made by multiplying it by them.

C-IX.9-2: That which is made of as many numbers as you wish in one, equals that which is made by their sum on it.

Campanus' proof of the first one is based on directly applying VII.5. For the proof of the second one he applied to the first the commutativity of the product, a property that appears in Campanus VII.17 (or Euclid's VII.16).

Campanus stressed in relation with C-IX.9–1, that "the first of the second" (i.e., Euclid's II.1) states the same thing but for lines. A similar statement appears in all the following commentaries, 4–12, with relation to each of the propositions II.2–10 respectively. The first three of these correspond to the distributivity-like properties II.2–4, and they are proved by direct application of the first two rules.

The important point to notice concerning these commentaries is that for the sake of their proof, Jordanus had had to start with a distributivity-like property for *multiplicities* (A-I.6). This provided the basis for proving other statements for arithmetic that are truly parallel to those of geometry, in the sense that they refer to a multiplication of *number by numbers*. Campanus, in turn, could base his proof directly on VII.5, which already handled the distributivity-like property of the multiplicities.

Campanus's version of the *Elements* had a decisive influence on the way that Euclid's treatise was read and understood over the following generations, particularly in relation with the issue of the relationship between arithmetic and geometry. This is of course also the case concerning distributivity-like properties. Of particular importance in this regard is the addition of purely arithmetic versions to Book IX. Readers of the Campanus version, or of any other work derived from it, would now have good grounds—and by all means better grounds than those of a reader of any previous treatise—for seeing these properties as inherently arising within the purely arithmetic realm, without any need for additional support coming from geometric considerations.

References

Allard, André. 1991. The arabic origins and development of Latin Algorisms in the twelfth century. *Arabic Sciences and Philosophy* 1 (2): 233–283. https://doi.org/10.1017/S0957423900001508.

Ambrosetti, Nadia. 2016. Algorithmic in the 12th century: the Carmen de Algorismo by Alexander de Villa Dei. In *History and Philosophy of Computing*, ed. Fabio Gadducci and Mirko Tavosanis, 71–86. Cham: Springer International Publishing.

Bar Chijja, Abraham. 1912. *Chibbur ha-meschicha weha Tischboreth. Lehrbuch der Geometrie des Abraham bar Chija hrsg. u. mit Anmerkungen versehen von Michael Guttmann.* 2 vols. Berlin: Schriften des Vereins Mekize Nirdamim.

Bar Hiia, Abraam, Miquel Guttmann, and José María Millás Vallicrosa. 1931. *Llibre de geometria Hibbur hameixihà uehatixbóret.* Barcelona: Editorial Alpha.

Boncompagni, Baldassare. 1856. *Opuscoli di Leonardo Pisano, pubblicati da Baldassare Boncompagni.* 2d ed. Firenze: Tipografia Galileiana. https://gallica.bnf.fr/ark:/12148/bpt6k8527691.

Busard, H.L.L. 2005. *Campanus of Novara and Euclid's Elements.* Stuttgart: Franz Steiner Verlag.

Corry, Leo. 2013. Geometry and arithmetic in the medieval traditions of Euclid's *Elements*: a view from Book II. *Archive for History of Exact Sciences* 67 (6): 637–705. https://doi.org/10.1007/s00407-013-0121-5.

Corry, Leo. 2016. Some distributivity-like results in the medieval arithmetic of Jordanus Nemorarius and Campanus de Novara. *Historia Mathematica* 43 (3): 310–331. https://doi.org/10.1016/j.hm.2016.06.001.

Curtze, Maximilian. 1897. *Petri Philomeni de Dacia In Algorismum vulgarem Johannis de Sacrobosco commentarius.* Hauniae: A.F. Host. http://archive.org/details/petriphilomenid00curtgoog.

Curtze, Maximilian. 1902. Der Liber Embadorum des Abraham Bar Chijja Savasorda in der Übersetzung des Plato von Tivoli. *Abhandlungen zur Geschichte der Mathematischen Wissenschaften* 12: 1–183.

Fibonacci, Leonardo. 1987. *The book of squares. An annotated translation into modern English by L. E. Sigler.* Boston: Academic Press.

Folkerts, Menso. 2001. Early texts on Hindu-Arabic calculation. *Science in Context* 14 (1 & 2): 13–38.

Folkerts, Menso. 2004. Leonardo Fibonacci's knowledge of Euclid's *Elements* and of other mathematical texts. *Bollettino di Storia della Scienze Matematiche* 24: 93–114.

Grant, Edward. 1974. *A source book in medieval science.* 2 vols. Cambridge, MA: Harvard University Press.

Hughes, Barnabas. 2008. *Fibonacci's de Practica Geometrie.* Sources and Studies in the History of Mathematics and Physical Sciences. New York: Springer.

Katz, Victor, Menso Folkerts, Barnabas Hughes, Roi Wagner, and J. Lenart Berggren, eds. 2016. *Sourcebook in the mathematics of medieval Europe and North Africa.* Princeton: Princeton University Press.

Lévy, Tony. 2001. Les débuts de la littérature mathématique hébraïque: la géométrie d'Abraham Bar Hiyya (XIe-XIIe s.). *Science in Context* 10 (03): 35–64.

McClenon, R.B. 1919. Leonardo of Pisa and his Liber Quadratorum. *The American Mathematical Monthly* 26 (1): 1–8. https://doi.org/10.1080/00029890.1919.11998476.

Moyon, Marc. 2012. "Algèbre & pratica geometriæ en occident médiéval latin : Abū Bakr, Fibonacci et Jean de Murs." https://hal.archives-ouvertes.fr/hal-00933015.

Rommevaux, Sabine. 2007. La similitude des équimultiples dans la définition de la proportion non continue de l'édition des *éléments* d'Euclide par campanus: une difficulté dans la réception de la théorie des proportions au moyen age. *Revue d'histoire des mathématiques* 13: 301–322.

Sarfatti, Gad B. 1968. *Mathematical terminology in Hebrew scientific literature of the Middle Ages* (מונחי מתמטיקה בספרות המדעית העברית של ימי הביניים). Jerusalem: Magnes Press.

Sesiano, Jacques. 2014. *The Liber Mahameleth*. Sources and Studies in the History of Mathematics and Physical Sciences. Cham: Springer International Publishing. http://link.springer.com/https://doi.org/10.1007/978-3-319-03940-4.

Simson, Robert. 1804. *The Elements of Euclid: also the book of Euclid's Data.* London: F. Wingrave.

Valleriani, Matteo, ed. 2020. *De sphaera of Johannes de Sacrobosco in the early modern period: the authors of the commentaries.* Berlin: Springer.

Vlasschaert, Anne-Marie., ed. 2010. *Le Liber Mahameleth: Edition critique et commentaires.* Stuttgart: Franz Steiner Verlag.

Chapter 5
Hebrew Mathematics

Abstract This chapter discusses how distributivity-like properties appeared in medieval Hebrew mathematical texts. It examines works by Gersonides, Alfonso de Valladolid, Eliezer Komtino and Elijah Mizrahi.

Keywords Distributive-like rules · The Hebrew medieval tradition · Gersonides · Alfonso de Valladolid · Eliezer Komtino · Elijah Mizrahi

This last chapter is devoted to examining how distributivity-like properties appeared and were discussed in medieval Hebrew mathematical texts, with a particular focus on the (relatively late) works of Gersonides, Alfonso de Valladolid, Eliezer Komtino and Elijah Mizrahi. Hebrew mathematical texts, and the history of their transmission, translation and assimilation, have attracted focused attention over the last three decades. This tradition has become known as "the Mathematical Bookshelf of the Medieval Hebrew Scholar" (Lévy 1997). In this section I rely strongly on recent scholarly work devoted to a variety of Hebrew texts that discuss works produced in Italy, Provence and the Byzantine world, while referring to a variety of sources, Hebrew and Arabic of course, but also more Latin and vernacular. Particularly useful is the ongoing web-based project "Mispar" (https://mispar.ethz.ch), created and maintained by Naomi Aradi and based at the ETH-Zurich, where many important Hebrew arithmetic treatises are directly available, partially with comments and in translation. I have consulted the site in detail in order to find out about the relevant texts to be discussed here, as well as to inspect others, no less interesting, but with no directly connected sections dealing with distributivity-like properties.

Any account such as intended here must start by referring to Bar Ḥiyya's *Ḥibbūr ha-meshīḥah we-ha-tishboret*. Above I indicated the great impact of the translated *Liber Embadorum* on the Latin tradition, and here I remark again that of the original Hebrew version, which had of course a marked influence on the Hebrew tradition to which the texts to be discussed here belong. Bar Ḥiyya is the earliest representative of a tradition involving Jewish sages who assimilated important parts of the Greco-Islamic scientific culture and published their own works in Hebrew, mainly between the 12th and 14th century. Such works were intended as a vehicle for transmitting Islamicate science to Jewish audiences not conversant with the Arabic language. They often comprised compilations and systematic presentations of previous work, but in many cases they also presented original ideas.

A characteristic of many of these texts is the noticeable effort made by the authors to stress their character as practical treatises that draw their contents from the "wisdom of the nations" (חכמת הגויים) but, that at the same time provide useful tools in the service of the teachings of the Jewish law. One should keep in mind, however, that in many Hebrew texts of the time, very much like in the Islamicate tradition, we find this kind of formulation used as a rhetorical device to justify the fact that a learned Jew is devoting his precious time and efforts away from the study of the Bible.

Bar Ḥiyya mixed in a purposeful and original way ideas taken from both geometry (חכמת השיעור) and arithmetic (חכמת המניין) and this had profound consequences on his entire presentation of results taken from the various books of Euclid (Corry 2013, 667–74). For instance, in his formulation of Euclid's II.4, Bar Ḥiyya speaks, as in Euclid's original, about a line that is divided at an arbitrary point and about the squares that are built on the resulting segments. But to this purely geometric formulation he added: "and I give you an example with numbers" (ואני נותן לך בזה דמיון מן המניין). The example is that of a line of length 12 divided into two segments of lengths 7 and 5. The square on the entire line (רבוע הקו כולו) is 144, which is equal to the square of 7 and the square of 5, and twice the product of 7 by 5 (כפל הרבוע ז' בה'). We will see a similar approach appearing in the much later texts discussed in this section.

5.1   Gersonides: *Sefer Maaseh Hoshev*

Sefer Maaseh Hoshev (ספר מעשה חושב), a title that can be translated as "The Work of the Calculator", dating from 1321, is a well-known arithmetic treatise written by Levy Ben Gerson (1288–1344), also known as Gersonides. In the introduction, Gersonides told his readers that he would assume a thorough knowledge of the arithmetical books of the *Elements*, and that he would not prove in his own text any of the results appearing in those books. Among the most basic results that he did prove there are some fully arithmetic versions of propositions from Book II, and specifically those embodying distributivity-like properties. The proofs do not contain new ideas, but it is interesting to take a closer look at the way in which they are presented here as purely arithmetic results with no trace of geometric origins or content. This can certainly be taken to represent the manner in which these kinds of results were already understood by the early fourteenth century (Corry 2013, 693–98).

For the sake of completeness, it is important also to stress that the distributivity rules on which we focus here appear in Gersonides' treatise together with all other basic rules of arithmetic which in his opinion provide the foundations for this

Fig. 5.1 Gersonides'
SMH-I.1

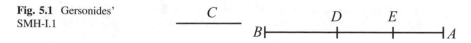

domain of knowledge. This includes the very definition of what is a multiplication of one number by another ("it counts every part of the first number as many times as there are ones in the second number"). More interestingly, he proves the associativity rule for the multiplication of three numbers, then for four numbers, and then, relying on a kind of original inductive argument, he generalizes this rule to any number of factors (Katz et al. 2016, 253–68).

The most relevant proposition to examine in the context of the present analysis is Gersonides' version of Euclid's II.1 (SMH-I.2), which reads as follows[1]:

SMH-I.2: When two numbers are given and one of them is divided into as many parts as we wish, the area of one of the numbers by the other equals the sum of the areas of each part of the one multiplied by the other.

The diagram he provides is reproduced in Fig. 5.1 (for convenience I use Latin letters instead of the Hebrew ones appearing in the original), and the proof reads as follows:

> For let the numbers *AB* and *C* be given, and let the number *AB* be divided into parts, *AE*, *ED*, *DB*. I say that the area of *AB* on *C* equals the area of *AE* on *C* and the area of *ED* on *C* and the area of *DB* on *C* taken together. Proof: the area of *AE* on *C* contains the number of units in *AE* as many times as *C*, the area of *ED* on *C* contains the number of units in *ED* as many times as *C*, and the area of *DB* on *C* contains the number of units in *DB* as many times as *C*. Therefore, all of them taken together contain the number of units of *AE*, *ED*, *DB*, taken together as many times as *C*. But the number of units of *AE*, *ED*, *DB*, taken together equals the number of units in *AB*.

The diagram is interesting because, on the face of it, it is similar to that of Heron (as well as others we saw above), but unlike with Heron, here the diagram actually accompanies an arithmetic kind of proof based on mere counting of units. Gersonides speaks here of, for instance, "the area of *AB* on *C*", and as in Heron's proof this is not represented in the diagram. Unlike with Heron, the result (even though it is called "area") is yet another number, but it is *not* represented by another line in the diagram as was usually the convention for arithmetic-type proofs, where all numbers as well as all products of numbers mentioned in the proof appear as lines in the diagram.

The next two propositions are just extension of SMH-I.2. Thus, SMH-I.3 is similar to Jordanus' A-I.11, whereas SMH-I.4 is parallel to Euclid's II.2:

> **SMH-I.3:** When two numbers are given and each of them is divided into as many parts as we wish, the area of one of the numbers by the other equals the sum of the areas of each part of the one multiplied by each and every part of the other number.

---

[1] Gersonides consistently uses the term "area" (שטח) to indicate products of numbers. But this choice derives from the limited mathematical lexicon available to him, and beyond it there is no hint of any kind of geometric thought involved here. All quotations are taken from (Lange 1909).

SMH-I.4: If any given number is divided into two numbers, the area of the number by any of its parts equals the area of one part by the other, together with the square on the part.

Both are proved by direct application of SMH-I.2, and they are used to prove some of the original propositions that Gersonides formulates in the book (Corry 2013, 693–698).

We thus see that, as with any other issue addressed in *Sefer Maaseh Hoshev*, Gersonides' treatment of distributivity-like properties was consistently arithmetic throughout. This was perhaps representative of the way in which most of his contemporaries had come to conceive of the entire issue by this time.

5.2 Alfonso de Valladolid: *Sefer Meyasher `Aqov*

Sefer Meyahser `Aqov (Rectifying the Curved—ספר מיישר עקוב) is another interesting mathematical treatise, nearly contemporaneous to *Maase Hoshev*, written in the first half of the fourteenth century. The author was the Jewish sage Abner de Burgos (c. 1270–c. 1347), who towards the end of his life converted to Christianity and adopted the name Alfonso de Valladolid. It is fortunate that a critical edition of the Hebrew manuscript, with English translation and commentaries, was recently published by Ruth Glasner and Avinoam Baraness (2021), since, as it happens, there are several sections in that treatise which are interestingly relevant to our discussion here. I will be citing from the text as reproduced in their book, and in addition I will be referring to diagrams included in the book and reproducing them. It is important to stress, however, that the diagrams do not appear in the original manuscript consulted by the authors. Rather, they were created by Baraness based on a careful reading of the text, and they prove to be very useful for understanding the proposition that they are meant to illustrate.

Sefer Meyahser `Aqov (*SMA*) presents an intriguing mixture of philosophical debates and technical, Euclid-like geometrical propositions. A main topic in the first two parts of the treatise is the attempt to legitimize the use of motion and superposition in geometry, as well as a discussion of the possible errors that may result from it, with a focused view on a possible solution of the problem of the quadrature of the circle. Then, Part III comprises a compilation of 33 geometrical propositions (8 of which are missing), meant to serve as basis for the main attack on the problem in Parts IV and V, which unfortunately are mostly missing.

Of the propositions of Part III, some embody well-known results, some embody lesser-known ones. Alongside some minor lemmas, we also find propositions which are highly interesting on their own, together with tools and ideas which are completely original, and worthy of attention. In particular, Alfonso recurrently relies on distributivity-like lemmas that are crucial for proving some of his main results. In this section I focus on those lemmas and in their usage by Alfonso.

Several of the original concepts adopted by Alfonso deal with triangles that are constructed on the basis of two given segments of straight lines. The Hebrew term is רבוי (literally "multiplicity"). Glasner and Baraness translated the term in this context as "expansion", and I will follow suit.

Thus, for instance, Alfonso introduced the following definitions:

- An expansion of a straight line by a straight line is the rectilinear right-angled triangle enclosed by the two lines.
- An expansion of a straight line by a straight line at a given angle is the rectilinear triangle enclosed by the two straight lines at the given angle.
- An expansion of a straight line by a straight line at a given angle expanded by a straight line is the pyramid whose base is the expansion of the two lines by one another and its height is the third line.

In their edition, Glasner and Baraness use the shorthand $a \otimes b$ to denote the expansion of straight line a by straight line b. For the sake of uniformity with previous sections, I will denote that expansion as $\mathrm{Tr}(a, b)$. As we will see now, there is a distributivity-like lemma that plays a crucial role in some of Alfonso's proofs and which can be formulated as follows:

$$\mathrm{Tr}(a+b,c) = \mathrm{Tr}(a,c) + \mathrm{Tr}(b,c).$$

To be sure, the lemma is not explicitly stated, much less justified in the text, and the question arises whether it may have been formulated in the parts of the manuscript which are not extant. I return to this point at the end of the section.

Let us take a closer look at how this appears in one of the interesting propositions of Part III, SMA-III.11. It is formulated as follows:

SMA III.11: If two straight lines are drawn from the two extremities of a diameter of a circle, intersecting [each other] and cutting the circumference of the semicircle, then the expansion of the part of the first line from the extremity of the diameter to the meeting point of the lines and the part of it which is a chord in the circle, together with the expansion of the second line's part from the [other] extremity of the diameter to the meeting point of the lines, and the part which is a chord in the circle, together they are equal to the expansion of the diameter of the circle by itself.

The statement is better understood by looking at the diagram reproduced in Fig. 5.2. AB is the diameter of circle $ABCD$, and lines AE, BE, intersecting at E, when extended, cut the circle's circumference at points C, D. EG is drawn perpendicularly to AB. The proposition then states that the expansion AC by AE, and the expansion of BD by BE taken together equal the expansion of AB by itself.

Schematically, the proposition can be rendered as:

$$\mathrm{Tr}(AC,AE) + \mathrm{Tr}(BE,BD) = \mathrm{Tr}(AB,AB).$$

Alfonso's proof is succinctly phrased. In order to fully understand it one must complete two steps that are only implicit. What the text says, can be rendered as follows:

Fig. 5.2 Alfonso's diagram
for *SMA*-III.11

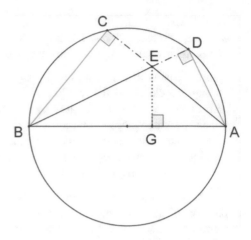

(n.1)   By construction: Δ *BGE* $\sim$ Δ *ABD*; $\measuredangle$ *BEG* = $\measuredangle$ *BAD*;
(n.2)   Hence: Tr(*BE*, *BD*) = Tr(*BG*, *AB*); [implicit step: explained below]
(n.3)   Similarly: Δ *AEG* $\sim$ Δ *BAC*; $\measuredangle$ *AEG* = $\measuredangle$ *ABC*;
(n.4)   Hence: Tr(*AE*, *AC*) = Tr(*AG*, *AB*); [implicit step: explained below]
(n.5)   From (n.2), (n.4): Tr(*BE*, *BD*) + Tr(*AC*, *AE*) = Tr(*BG*, *BA*) + Tr(*AG*, *AB*);
(n.6)   Hence: Tr(*AC*, *AE*) + Tr(*BE, BD*) = Tr(*AB, AB*). [the distributive step!!]

QED

The implicit steps are as follows:

(o.1)   From (n.1): *BE*:*BG* :: *BA*:*BD*;
(o.2)   Hence: Sq(*BE*, *BD*) = Sq(*BG*, *AB*); [by Euclid's VI.6]
(o.3)   Hence: Tr(*BE*, *BD*) = Tr(*BG*, *AB*);
(o.4)   Likewise: *AE*:*AG* :: *AB*:*AC*;
(o.5)   Hence: Tr(*AC*, *AE*) = Tr(*AG*, *AB*). [by Euclid's VI.6]

Notice that if triangle *AEB* is right-angled at *E*, then obviously what we have is the Pythagorean Theorem. Indeed, Alfonso indicates explicitly that this proposition is a generalization of that theorem. An interesting issue that might arise here, but that Alfonso does not address—as Glasner and Baraness rightly point out in their account—concerns the case of the angle at *E* being acute, with *E* lying outside the circle, as represented in Fig. 5.3.

Indeed, as pointed out above, in Euclid's *Elements* the proofs of the Pythagorean Theorem, I.47, as well as of some generalizations of it, raise interesting questions concerning distributivity-like properties, and the question of visible versus invisible constructions. Alfonso's proof of SMA-III.11 is a classic example of an invisible construction that relies on a distributivity-like property that is not explicitly stated, and that needs to be tacitly assumed in order to complete the argument. In the *Elements*, Euclid had generalized the Pythagorean Theorem for both the acute- and obtuse-angled triangles (II.12, II.13), and both proofs relied explicitly on

Fig. 5.3 A diagram
representing a variant of
SMA-III.11, not considered in
the treatise

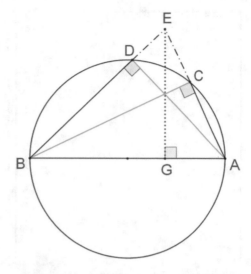

distributive-like propositions (respectively: II.4, II.7). But remarkably, the "extensions" involved in SMA-III.11 concern triangles built on lines cut at an arbitrary point, and in this sense they run parallel to all the basic theorems of Book II of the *Elements*. Similar is the situation in many other of the somewhat haphazard collection of geometric propositions that conform this part of the treatise (especially SMA III.10–23). This kind of closeness with the propositions of Book II cannot be coincidence and it is more than likely that Alfonso intended his propositions as a broadening of the basic tools originally developed in Book II of the *Elements*.

Let us consider now how this works with a second example, namely, SMA III.15, which is formulated as follows:

> **SMA III.15:** In every rectilinear right-angled triangle circumscribing a circle, the expansion of the two parts of the hypotenuse, as divided by the tangent of the circle, by one another is equal to half of the triangle.

The diagram represented in Fig. 5.4 describes well the situation: *ABC* is a right-angled triangle, with *AC* its hypotenuse. *ABC* circumscribes a circle *KHI*, *with* points *K, I, H* being tangent points of the circle and the sides. *GH, GI, GK* are radiuses and they are perpendicular to the sides of the triangle. Also, *GH = GK = GI = HB = BI* and *AK = AI, BI = BH, CH = CK*. Moreover, as the triangle circumscribe the circle, the three lines *GA, GB, GC* bisect the angles of the triangle. The proposition then states that the expansion of *AK* by *KC* is equal to half the triangle.

Schematically, this formulation can be rendered as follows:

$$\mathrm{Tr}(AK, CK) = \tfrac{1}{2}\,\mathrm{Tr}(AB, BC).$$

So, here we have once again a property that concerns a figure built on two segments cut at a certain point, out of a given line. And assuming once again the

Fig. 5.4 Alfonso's diagram
for *SMA*-III.15

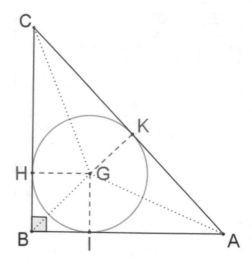

distributivity-like property of expansions, Alfonso's proof is quite succinct and densely phrased. It can be schematically rendered as follows:

(p.1)   Tr(AB, BC) = Tr(AI + BI, BH + HC);

(p.2)   Tr(AB, BC) = Tr(AI, BH + HC) + Tr(BI, BH + HC); [distributivity]

(p.3)   Tr(AB, BC) = Tr(AI, BI + HC) + Tr(BI, BH + HC); [since BH = BI]

(p.4)   Tr(AB, BC) = Tr(AI, CH) + Tr(BI, AI) + Tr(BI, BH + HC);

(p.5)   Tr(AB, BC) = Tr(AI, CH) + Tr(BI, AI + BH + CH); [distributivity]

(p.6)   Tr(AB, BC) = Tr(GI, AI + BH + CH) + Tr(AK, CK). [AI = AK, CH = CK, GI = BI]

(p.7)   On the other hand: Δ ABC = Δ ABG + Δ ACG + Δ BCG;

(p.8)   Hence: Tr(AB, BC) = Tr(AB, GI) + Tr(AC, KG) + Tr(BC, GH);

(p.9)   Hence: Tr(AB, BC) = Tr(AB + AC + BC,GI); [distributivity]

(p.10) Tr(AB, BC) = Tr(BH + CH + BI + AI + AK + CK, GI);

(p.11) Tr(AB, BC) = Tr($2 \cdot AI$ + $2 \cdot BII$ + $2 \cdot CII$, GI) = $2 \cdot$Tr(AI + BH + CH, GI).

(p.12) Finally, by (p.5), (p.10): ½·Tr(AB, BC) = Tr(AK, CK).

QED

The way in which these two proofs, as well as some additional ones in Part III of *SMA*, rely on a distributivity-like kind argument is evident and prominent. The question remains open whether the argument, or anything similar to it, may have been explicitly formulated in parts of the manuscript which are not extant. Glasner and Baraness mention this as a possibility, alongside the alternative conjecture that Alfonso simply took such distributivity-like properties for granted. For lack of direct evidence to decide, I conjecture that it is more reasonable to believe that the latter is the case. The pervasive use of distributivity-like properties appearing in earlier texts, some of which I mentioned above and some of which Alfonso likely read as part of his mathematical background, make that view plausible.

Another interesting issue that Glasner and Baraness raise (p. 134), but left unanswered for lack of direct evidence, concerns the question why did Alfonso prefer to define the basic geometrical operation of his technical presentation—"expansion" of two lines—as the right-angled triangle enclosed by the two lines, rather than as the rectangle enclosed by them. Also puzzling is the fact that he preferred to define the three-dimensional expansion as the pyramid erected on the plane expansion, rather than as the right prism. Glasner and Baraness speculate that, perhaps, "thinking of area in terms of triangles could have reflected Plato's conception of the triangle as the basic component of reality or Archimedes' describing the area of a circle to a triangle." This is a thought-provoking possibility but I am not sure whether there are good reasons to endorse it. It remains, I believe, as an open question to be further investigated.

5.3 Eliezer Komtino and Elijah Mizrahi: *Arithmetic*

Sefer ha-Mispar (The Book of Number—ספר המספר) is a rich and interesting arithmetic treatise written in the second half of the fifteenth century by Elijah Mizrahi (ca. 1450–1526). Also here we find interesting ideas which are relevant for our account. Mizrahi was the most important rabbinical authority of his generation in Constantinople and in the Ottoman Empire at large. He had a keen interest in scientific issues and he wrote, among other things, a commentary on Ptolemy's *Almagest*. Several complete manuscripts of *Sefer ha-Mispar* are preserved, which attest to the broad audiences it reached. A first printed edition appeared right after Mizrahi's death and it was also partially translated into Latin (Segev 2010, 2016).

Mizrahi's teacher on secular matters was Mordekhai ben Eliezer Komtino (1402–1482), a prominent Jewish sage active in Constantinople at a time of momentous historical importance. In 1453 the Ottomans conquered the city, and the Judeo Byzantine communities in the empire came to absorb a stream of exiles from the Iberian Peninsula who brought with them influential intellectual traditions, including those of marked scientific character. Komtino's treatise, *Sefer ha-Ḥeshbon we ha-Middot* (Book on Computation and Mensuration—ספר החשבון והמידות), covered a similar territory than Mizrahi's *Sefer ha-Mispar*, but in a much less elaborated manner. Although it was less read than Mizrahi's treatise, Komtino's *Sefer ha-Ḥeshbon we ha-Middot* set the stage for it. Its arithmetic sections are worthy of comment here.

Though the natural and more immediate context for analyzing the texts of both Komtino and Mizrahi is that of Jewish mathematics, there is a second intellectual context that needs to be taken into account as well, and at least mentioned in passing here. This the context of the Byzantine scholarly world—with Byzantine mathematics being part of it—particularly in the fourteenth and fifteenth centuries. The Byzantine world of knowledge has been the subject of recent historical research (see, e.g., Lazaris 2020), which provides an illuminating context for understanding the place of Byzantine mathematical texts, their sources and achievements. Of particular relevance for our account here are the contributions of George Pachymeres (1242–c. 1310) and of the Basilian monk Barlaam de Seminara (ca. 1290–1348), both of whom came up with arithmetic versions of Book II of the *Elements*. Each of their texts may have served to shape Komtino's ideas, though we do not have direct evidence for that, as is typically the case when it comes to Byzantine scientific culture (Acerbi 2020).

Like many Hebrew mathematical treatises, Komtino's opened with an explanation of the importance of a solid knowledge of both arithmetic and geometry, as they allow for a fair interaction between people, both in terms of commerce and of "compassion towards the needy." Invoking the authority of Nichomacus, Komtino stressed that the arithmetic part of his book (חכמת החשבון) precedes the geometric one (חכמת המידות) because of the precedence of arithmetic in nature.[2]

The arithmetic sections of Komtino's treatise add up to a succinct discussion of the topics typically handled in earlier *algorismus* tracts, written in either Arabic, Latin or Hebrew. Like them, *Sefer ha-Ḥeshbon we ha-Middot* explains how to perform the basic operations with numbers written in the Hindu-Arabic system (not including separate sections for halving or doubling) and applies them to solving word problems. Komtino relied on various sources, including Abraham bar Ḥiyya's *Hibbur*, and the *Arithmetica* of Nichomacus (Schub 1932; Silberberg 1905). Also the influence of another earlier Hebrew treatise, *Sefer ha-Mispar* by Abraham ibn Ezra (1108–1167), is clearly felt in the text. In the geometric sections of the treatise, which I will not discuss here, the influence of the Heronian and pseudo-Heronian metrological tradition is strongly manifest, as recently shown in (Lévy and Vitrac 2018).

Komtino's section on multiplication discusses some methods that facilitate calculating in specific cases. Such shortcut methods were part of a long standing tradition, and in particular they are discussed in ibn Ezra's *Sefer ha-Mispar*. Some of them rely on distributive-like principles that are used but not made explicit, much less justified. Thus, for instance, Komtino first explains how to square a number that is divisible by five, say sixty. The steps to be followed are these:[3]

> The fifth of the number is 12. You multiply by itself to obtain 144. You move it to the next rank [i.e., you multiply by 10], and it turns into one thousand four hundred and forty, which you multiply by 2 and half and it turns into 3 thousand and 360, which is the result.

This is a simple case that does not involve any distributive-like idea, and that can be rendered as:

$$\left(\frac{n}{5}\right)^2 \cdot 10 \cdot 2.5 = n^2.$$

Slightly more complicated, however, is the case where the number to be squared is not divisible by five. In that case, one chooses the number closest to the given

---

[2] Both Komtino's and Mizrahi's texts are available at the Mispar website, http://mispar.ethz.ch. In the footnotes below I have cited some specific passages as they appear there. Segev (2010) also contains a transcription of Mizrahi's text, as well as many useful commentaries and clarifications.

[3] וחמישיתו י"ב כפלנום על עצמם נהיו קמ"ד העלנו אותם במדרגה הבאה הסמוכה להם ונהיו אלף וארבע מאות וארבעים כפלנום על ב' וחצי ונהיו ג' אלפים ת"ר וזהו המבוקש.

one, either preceding or subsequent, but with a difference not larger than two. If, for example, the number divisible by five is the one immediately preceding the given number, then we square this number by the method described above, and then to the square thus obtained we add the given number together with the one that precedes it, and that will be the desired result. If, on the other hand, the number divisible by five precedes the given number by two, then we square it by the method described above, and then to the square thus obtained we add the double of adding the given number to preceding one, and that will be the desired result.[4] Komtino also explained in similar terms how to calculate when the number which is divisible by five follows the given number, by either one or two, and then he worked out the examples of squaring 11, 12, 13 and 14.

Now, when we render these rules algebraically, it is easy to see the rationale that underlies their justification, and this rationale is based on distributivity-like principles. Indeed, if the given number is n and the one which is divisible by five is $n - 1$, then Komtino's rule stipulates that

CO-1: $n^2 = (n - 1)^2 + n + (n - 1)$.

Retrospectively understood, this result can be seen to be correct on the basis that $(n - 1)^2 = n^2 - 2n + 1$. Likewise, when the number divisible by five is $n - 2$, then Komtino's rule stipulates that

CO-2: $n^2 = (n - 2)^2 + 2 \cdot [n + (n - 2)]$.

Also in this case, we can retrospectively realize that it is correct on the basis that $(n - 2)^2 = n^2 - 4n + 4$.

Another related, distributive-like rule which Komtino provides is equivalent to the rule $(a - b) \cdot (a + b) = a^2 - b^2$. It is important to stress once again that, like in the Hebrew and Islamicate traditions in general that stand in the background to this work, this rule is not intended as an algebraic identity involving *unknown quantities*, but as an arithmetic one involving *arbitrary numbers*. He formulated this as an extension of the previous rules on squaring:[5]

> **CO-3:** If you wish to multiply two numbers that deviate from the rule in that one of them falls short from a number and the second one exceeds it (by the same value), you can square by the rule the value that exceeds and falls short and subtract it from the square of the said number and this is the desired result.

This is followed by the example of the product of 25 by 35, which is calculated by squaring 5 and then subtracting it from the square of 30. This example is one of

---

[4] ואם היה המספר אשר יש לו חמישית הקרוב אל מספרך קודם ממספרך באחד תחשבהו בדרך שהזכרתי ראשונה על המספר שיש לו חמישית אחר תחבר מספרך עם המספר שיש לו החמישית ותוסיפהו על הסך והוא המבוקש. ואם היה ההפרש שנים תכפול מספרך אחר שתחברהו עם המספר שיש לו חמישית עם שנים ואחר תוסיפנו על הסך והוא המבוקש. אמנם אם המספר שיש לו החמישית הוא אחר מספרך תחשוב חשבונך על המספר שיש לו החמישית אחר תחבר מספרך עם המספר פעם אחת אם ההפרש אחד או תכפול אותו עם ב' אחר שחברתו אם ההפרש שנים ותגרעהו מהסך והנשאר הוא המבוקש.

[5] אם תרצה לכפול מספר על מספר שהם רחוקים מהכלל בשווי האחד חסר ממנו והאחר מוסיף עליו אתה יכול לכפול הכלל על עצמו אחר תרבע המספר היתר והחסר ותגרעהו מחשבון הכלל וההוה הוא המבוקש.

the places where the direct connection of the treatise with the existing Hebrew traditions can be noticed, as a similar formulation of the rule, and a similar example appear in ibn Ezra's *Sefer ha-Mispar*.[6]

To conclude this section, Komtino briefly added yet another rule that involves a distributivity-like idea intended as a shortcut for facilitating multiplication. Indeed, if in general the product of two two-digit numbers requires four individual multiplications, it is possible to reduce this to only three when—in both cases—the digit in the tens is one. For example, if we want to multiply 13 by 14, we can follow these steps: (1) add the units, 3 and 4, to obtain 7; (2) multiply 10 by 7, to obtain 70; (3) multiply 10 by 10 to obtain 100; (4) multiply 3 by 4 to obtain 12; (5) the result is obtained by adding 70, 100 and 12.[7] Three products in all.

In modern algebraic terms the rule followed in this case is simply:

CO-4: $(10 + a) \cdot (10 + b) = 100 + 10 \cdot (a + b) + (a \cdot b)$.

Komtino, to be sure, only provides the above example and he neither formulates a general rule nor provides a justification for the steps followed. More importantly, he does not go as far as formulating—much less justifying—a simple, but more general distributive rule, such as:

CO-4': $(10p + a) \cdot (10p + b) = (10p)^2 + 10p \cdot (a + b) + (a \cdot b)$.

Still, after having discussed the four rules above, with examples but without proofs, Komtino proceeded immediately, without any further explanation, to introduce the first ten propositions of Euclid's Book II. He explained that it is important for the reader to learn these properties of numbers (דברים מעניני המספר) since they help calculating more quickly (כדי שתהיה זריז בעניינים אלו). And the reader can only guess that the propositions may be seen as a justification for the rules just presented, although the former and the latter are not directly connected in any explicitly manner in the text.

Methods for mental calculations or verification of results do appear in some earlier Hebrew treatises. For instance, an anonymous manuscript dating from the late twelfth or early thirteenth century, and clearly influenced by works of bar Ḥiyya and Ibn Ezra, is devoted solely to such methods of mental computations (Aradi 2013).[8] At the same time, texts in the Islamicate tradition with a main focus on proto-algebraic techniques for problem solving, like those of al-Khwārizmī or al-Uqlīdisī, also discussed such methods, sometimes while explicitly indicating their connections with the Euclidean propositions of Book II (Segev 2010, 73). Both kinds of sources may have stood in the background to Komtino's presentation.

---

[6] See the examples in the transcription available at: https://mispar.ethz.ch/wiki//_ספר_המספר_אברהם_אבן_עזרא#Shortcuts.

[7] צריך אתה לדעת שאם היה כללם היה כללם אחד יספיק לכפלם שלש פעמים, כגון הרוצה לכפול י"ג על י"ד והנה כלל שניהם עשרה תחבר שני הפרטים ויהיו ז' ותכפול ז' פעמים י' נהיו ע' ויّ פעמים י' הרי ק' וג' פעמים ד' הרי י"ב והעולה קפ"ב וזהו המבוקש.

[8] A full transcription of the text is available at: https://mispar.ethz.ch/wiki/Anonymous.

In Komtino's treatise, each of the ten propositions of Book II is formulated as a property of operations with numbers, and, like the calculation rules that he taught, they appear without proof but illustrated with examples. Komtino may have learnt about the propositions in this form and context from al-Nayrīzī's commentary to Heron (the later only gives numerical examples for the first five propositions) or from some other Islamicate source or a translation thereof into Hebrew.[9] He does not mention any of these explicitly. At any rate, the first three propositions, at least, are clearly presented as distributive laws for the multiplication of numbers, and Komtino stressed their usefulness as aids for easier calculation. Thus, for instance, the proposition corresponding to II.3 reads as follows:[10]

> For any number divided into two parts as you wish, the product of the whole number by any of its two parts is equal to the product of the one part by the other plus the square of the part by which you multiplied the whole number. For example, divide the number ten into two parts, three and seven. Multiply it by three to obtain 30, which is equal to multiplying 3 by seven, which is 21, and the square of 3, which is 9, which is the part that we multiplied by itself. And likewise, if we multiplied 7, which is one of the parts, by ten we obtain seventy and this is equal to the product of 3 by 7, which is 21, and the square of 7 which is 49, which is the part that we multiplied, and which taken together make seventy.

Also Elijah Mizrahi discusses, in the section on multiplication of his *Sefer ha-Mispar*, procedures for checking the results of multiplications as well as methods of mental multiplication that can be applied in certain specific cases. In the basic organizational scheme of the treatise, Mizrahi first presents the algorithms and examples concerning each of the topics discussed, and only then he provides more detailed explanations and justifications. Mizrahi's treatise is quite unique in this regard, in that it provides such explanations in almost all of the sections. This is something rarely found in other similar treatises, and in comparison with Komtino, for example, the contrast is quite remarkable.

Procedures for checking multiplications, such as casting out nines, appeared previously in texts such as al-Khārizmī's *Treatise on Arithmetic* and, in fact, they were well-known even earlier than that (Folkerts 2001, 25–37). The source typically mentioned in Islamicate texts for methods of mental multiplication is Euclid's Book II (Saidan 1978, 384), but in fact the actual source may have been Heron's version or Nichomacus' texts. One way or another, in this part of Mizrahi's treatise we find again, like in Komtino's, interesting instances of relying on distributive-like properties.

Like Komtino, Mizrahi taught shortcuts to square numbers that are divisible by five, and then numbers that either fall short or exceed number that are divisible by five. However, he also taught similar methods that work for three, four and seven.

---

[9] See (Segev 2010, 81–87) for a more detailed discussion of the possible sources of Komtino's and Mizrahi's use of Euclid's Book II.

[10] כל מספר שחלקת אותו לשני חלקים איך שקרה הנה כפל המספר כולו על אחד משני חלקיו איזה שיהיה שוה לכפל החלק האחד על השני ולמרובע החלק משניהם אשר כפלת על כל המספר. דמיון המספר עשרה חלקנוהו לשני חלקים על שלשה ושבעה. כפלנו השלשה עמו והיו ל' וזה שוה לכפל הג' על השבעה שהם כ"א ולמרובע הג' שהם ט' שהוא החלק אשר כפלנו. גם ככה אם היינו כופלים הז' שהוא החלק האחד על העשרה היו שבעים וזה שוה לכפל הג' על הז' שהם כ"א ולמרובע הז' שהם מ"ט שהוא החלק אשר כפלנו ושניהם שבעים.

In fact, he expressed his surprise about the fact that "the ancients" (he didn't specify whom he was referring to) focused on the method with five, given that the basic idea is the same one in all cases. At the same he indicated that in applying these methods, the smaller the number the better, because then there are less exceptional cases to be handled separately. Thus, given a number that is divisible by three, in order to square it, "take its third part, multiply it by itself, move to the next rank [i.e., multiply it by ten], and subtract from it the square of its third."[11] Symbolically, this is equivalent to:

$$n^2 = \left[10 \cdot \left(\frac{1}{3}n \right)^2 \right] - \left(\frac{1}{3}n \right)^2.$$

If we approach the problem when considering the case of three, then there are only two cases to be handled for numbers that are not divisible by it, namely numbers that either fall short by one or exceed by one a number which is divisible by three. For instance, if the given number exceeds by one, Mizrahi provides the following method of squaring the given number: square by the above method the number divisible by three that the given number exceeds, and add to it the given number and the number that is divisible by three.[12] Symbolically, this amounts to saying that

$$(n+1)^2 = n^2 + (n+1) + n,$$

or equivalently, to assuming the simple, distributive-like relation

$$(n+1)^2 = n^2 + 2n + 1,$$

and then to apply it to a case where it is easy to calculate n^2.

After explaining the method without providing any justification for it, Mizrahi introduced the first ten propositions of Euclid's Book II in arithmetic formulation, without proofs but with examples for each case. He did not connect the propositions explicitly with the rules previously introduced, but he indicated to the reader that since he had seen "such beautiful and very useful things in these kinds of ideas mentioned by the learned Euclid in his book", he thought that it would be adequate to present them here "for the great benefit of the students."[13]

But then, later on in the treatise, Mizrahi provided explanations for all the methods of multiplication that he had introduced in the earlier sections while implying, in a somewhat general way, that the justification for the various methods, which are distributive-like in essence, arises from the propositions of Book II, seen

---

[11] קח שלישיתו והכהו עם עצמו והעולה העולה מדרגה אחת גבוהה ממדרגתו וגרע מרובע השלישית ממנו והנשאר הוא העולה מהכאת המספרים ההם בעצמם.

[12] ולהיות שראיתי דברים יפים מועילים מאד בזה המין הזכירם החכם אקלידס בספרו, ראיתי להביאם הנה למה שתגיע תועלת גדולה לתלמידים.

[13] בזה המין הזכירם החכם אקלידס בספרו ראיתי להביאם הנה למה שתגיע תועלת גדולה לתלמידים.

as numerical relations involving the numbers that are multiplied and their differences. But interestingly enough, Mizrahi opened this explanatory section formulating again Euclid's II.1 in *geometrical terms* (i.e. lines and areas) and stressing that the proposition is also valid for numbers because *the proof is also valid for numbers* (כי המופת צודק בהם)! This brings us back to the question of the source or sources that Mizrahi was relying upon. Obviously this was not Euclid's original one, or any version that followed it closely, because, as already explained, Euclid's proof cannot be naturally translated into arithmetic, as opposed to Heron's or any proof similar to it, which, because of its *operational* character, can indeed be translated.

In closing this section, I would like to mention yet another Hebrew arithmetical text, *Qeṣat mi-'Inyanei Ḥokmat ha-Mispar* (Some issues of arithmetic – קצת מעייני חוכמת המספר). This is an anonymous treatise that survived in a single manuscript dated to 1572, but there are some indications suggesting that it was initially written at a much earlier date. It makes use of many Arabic terms, which led Steinschneider to wonder whether the work is a translation of an Arabic text or even a lost part of Abraham bar Ḥiyya's encyclopedia. The last option can most likely be ruled out today since the arithmetical part of the encyclopedia is already restored. Moreover, the terminology used in the text is different from Bar Ḥiyya's terminology and the character of the textbook does not coincide with the encyclopedic nature of Bar Ḥiyya's work. The book is incomplete and its original order of discussion is uncertain as a few short passages on geometric issues are interpolated in the middle of the text (Aradi 2015).

The existing parts of this text include a discussion on sums and progressions, basic arithmetic operations with fractions, square and cubic roots of integers and of fractions, divisors of integers, square numbers, restoration and conversion, geometric shapes, and Euclidean propositions.[14] The Euclidean propositions discussed in a special chapter are phrased in an arithmetical language, and also proved and illustrated arithmetically. Most of the issues studied in the book are treated also in *Sefer ha-Mispar* by Elijah Mizrahi, illustrated by the same numerical examples, which are presented in the same order. The Hebrew wording on the other hand is sometimes different. It is quite clear therefore, either that both texts relied on a shared source or that the anonymous book discussed here itself was one of Mizrahi's sources (the opposite possibility is less likely, since Mizrahi does not use Arabic terms).

References

Acerbi, Fabio. 2020. Logistic, arithmetic, harmonic theory, geometry, metrology, optics and mechanics. In *A companion to Byzantine science*, ed. Stavros Lazaris, 105–59. Brill's Companions to the Byzantine World 6. Leiden-Boston: Brill. https://brill.com/view/title/55705.

Aradi, Naomi. 2013. An unknown medieval hebrew anonymous treatise on arithmetic. *Aleph* 13 (2): 235–309.

---

[14] The text is available at https://mispar.ethz.ch/wiki/קצת_מעייני_חכמת_המספר.

Aradi, Naomi. 2015. The arithmetic of the Jews in the Middle Ages: introduction and selected samples [in Hebrew]. PhD Thesis, Hebrew University of Jerusalem.

Corry, Leo. 2013. Geometry and arithmetic in the medieval traditions of Euclid's *Elements*: a view from Book II. *Archive for History of Exact Sciences* 67 (6): 637–705.

Folkerts, Menso. 2001. Early texts on Hindu-Arabic calculation. *Science in Context* 14 (1 & 2): 13–38.

Glasner, Ruth, and Avinoam Baraness. 2021. *Alfonso's Rectifying the Curved. A fourteenth-century Hebrew geometrical-philosophical treatise*. Sources and Studies in the History of Mathematics and Physical Sciences. Cham: Springer.

Katz, Victor, Menso Folkerts, Barnabas Hughes, Roi Wagner, and J. Lenart Berggren, eds. 2016. *Sourcebook in the mathematics of medieval Europe and North Africa*. Princeton: Princeton University Press.

Lange, Gerson. 1909. *Sefer Maassei Choscheb. Die Praxis des Rechners: ein hebräisch-arithmetisches Werk Levi ben Gerschom aus dem Jahre 1321*. Frankfurt a. M.: Buchdruckerei L. Golde.

Lazaris, Stavros, ed. 2020. A companion to Byzantine science. A companion to Byzantine science. *Brill's Companions to the Byzantine World* 6. Leiden-Boston: Brill. https://brill.com/view/title/55705.

Lévy, Tony. 1997. The establishment of the mathematical bookshelf of the Medieval Hebrew scholar: translations and translators. *Science in Context* 10 (03): 431–51.

Lévy, Tony, and Bernard Vitrac. 2018. Hero of Alexandria and Mordekhai Komtino: The encounter between mathematics in Hebrew and the Greek metrological corpus in fifteenth-century Constantinople. *Aleph: Historical Studies in Science and Judaism* 18 (2): 181–262.

Saidan, Ahmad Salim. 1978. *The arithmetic of Al-Uqlīdisī: The story of Hindu-Arabic arithmetic as told in Kitāb al-Fuṣūl Fī al-Ḥisāb al-Hindī*. Dordrecht: Springer, Netherlands.

Schub, Pincus. 1932. A mathematical text by Mordecai Comtino (Constantinople, XV Century). *Isis* 17 (1): 54–70.

Segev, Stela. 2010. The book of the number by Elijah Mizrahi. A textbook from the 15th century. PhD Thesis, Hebrew University of Jerusalem.

Segev, Stela. 2016. Elijah Mizrahi, Sefer Hamispar (The Book of Number). In *Sourcebook in the mathematics of medieval Europe and North Africa*, ed. Victor Katz, et al., 244–53. Princeton: Princeton University Press.

Silberberg, Moritz. 1905. Ein Handschriftliches Hebräisch-Mathematisches Werk Des Mordecai Comtino. *Jahrbuch Der Jüdisch-Literarischen Gesellschaft* 3: 277–92.

Chapter 6
Concluding Remarks

Abstract Important changes affected the Euclidean traditions in the passage from manuscript to printed versions. This also had an impact on the role of disrtibutive-like ideas.

Keyword Distributive-like rules · Printed editions · Euclidean traditions · Early modern mathematics

The passage from the medieval traditions of the *Elements* and its derivatives to the print culture of early modern mathematics brought with it many remarkable changes. The first printed version of the *Elements*, usually known as the Erhard Ratdolt edition, appeared in 1482 in Venice. It was based on Campanus' translation. The first published translation into Latin from a Greek text of the *Elements* appeared in 1505, also in Venice. The translator, Bartolomeo Zamberti (c. 1474–after 1539) was an influential humanist. He supported a view, commonly associated with Proclus, that stressed the "marvelous" nature and unity of the *Elements*. At the same time, he was also instrumental in popularizing a view according to which, the choice of definitions, postulates, and propositions in the *Elements* were Euclid's, whereas the demonstrations were Theon's. This alternative view opened the way to editions in which the Euclidean text was altered, summarized, divided, or published without the proofs.

In 1516 Jacques Lefèvre d'Étaples attempted to reconcile the numerous discrepancies between existing translations of the *Elements* and came up with a unified edition in which the Radtold's and the Zamberti's text appeared side-by-side. Eventually these two traditions, one stemming from Arabic sources (via Campanus) and one from Greek sources (via Zamberti), merged in the last third of the sixteenth century. In 1572 the edition of Federico Commandino (1509–1575), based on this new merging of traditions, appeared in print, marking an important milestone in the process of assimilation of the *Elements* as it came to be considered and understood in Europe. Many other treatises related in various ways to the *Elements* also started to appear in print, and its contents was influenced by these processes.

It is evident that also the issue of distributivity-like situations and the eventual systematic codification of distributivity rules along the sixteenth century, while attributing them foundational roles in arithmetic and algebra such as seen in the example of Viète mentioned in the introductory section, was part and parcel of these broader processes affecting the Euclidean tradition.

Alongside the changing approaches in the *Elements* to some of the central conceptions underlying the treatment of geometry and arithmetic, continuous and discrete quantities, and the interrelations among them, new and vigorous trends of symbolic algebraic techniques began to attract increased attention and were gradually incorporated into the text of the new version of the treatise. These new algebraic trends fitted in a relaxed manner into, and offered a natural continuation of, the kind of generalized arithmetic that Campanus, especially under the influence of Jordanus, had already made appear as the fundamental way to look at many results of Book II and elsewhere in the treatise.

The above account of the ways in which distributivity-like properties are formulated and proved in a variety of works stemming from three different mathematical cultures affords an interesting, and I believe quite original, perspective on fundamental medieval mathematical conceptions. The absence of a systematic foundation in the arithmetical books of the *Elements*, parallel to that of the geometrical ones, gave rise to a variety of ways to present handle this type of knowledge. Contemporary conceptions about the relationship between discrete and continuous magnitudes and the legitimate ways to handle them had become in these cultures much more flexible than those typical of the *Elements* in its original formulations, and the way of handling distributive-like properties in a variety of treatises bear witness to the more flexible approaches to the idea of number that had started to develop by then.

Within the renaissance traditions of the *Elements*, distributivity-like situations came to be understood in ways that went well beyond the purely geometric one that had informed Euclid's original conception. A detailed analysis of this topic is well beyond the scope of the present account and it will be left for additional enquiry at a future opportunity.

Printed in the United States
by Baker & Taylor Publisher Services